CHALLENGES OF CLIMATE CHANGE – Indian Horticulture

H.P. Singh
J.P. Singh
S.S. Lal

CENTRAL POTATO RESEARCH INSTITUTE
SHIMLA

WESTVILLE PUBLISHING HOUSE
NEW DELHI

H.P. Singh, J.P. Singh and S.S. Lal (2010) Challenges of Climate Change – Indian Horticulture, pp 224, Westville Publishing House, New Delhi

First Edition 2010

ISBN 978-81-85873-55-8

Published by Mrinal Goel, Westville Publishing House
47, B-5, Paschim Vihar, New Delhi – 110 063
Tel: 011-25284742, Fax: 011-25267469
Email: westville_2002@yahoo.co.in
Printed at Salasar Imaging Systems, New Delhi

Foreword

Global warming and climate change is being perceived as greatest threat to mankind. It may affect rainfall patterns, increase frequency of extreme events of drought, frost and flooding. It will also affect output of agriculture and allied sectors. Therefore, it has attracted the attention of various stakeholders for possible mitigation and adaptation measures. Most of the recent deliberations on the issue have concluded that climate change is real, which needs public intervention and also preparedness to face the same effectively.

Global climate change is expected to affect agricultural/horticultural crops through its direct and indirect effects. Scientific evidence suggests a positive effect of increase in atmospheric carbon dioxide in C_3 photosynthetic pathway. However, expected positive effects are likely to be counteracted due to increase in the temperature. Sensitivity of individual crop to temperature depends on inherent tolerance and growing habits. The net effect of climate change on horticultural crops will depend upon interaction effect of rise in temperature and CO_2 concentration in atmosphere on crops and pathogens. Before resorting to any adaptation options, a detailed investigation on the impact of climate change on horticultural crops is, therefore, essential.

Developing strategies and tools to comprehensively assess the impact of climate change and evolve possible adaptation measures in horticultural crops is less understood. Therefore, it is imperative to understand the effect of such changes on growth and development, quality of crops and the innovations that can help in adaptation and mitigation of climate change effects.

I am happy to know that scientists in the ICAR have effectively deliberated on the impact of climate change on Indian horticulture. Rise in temperature shall reduce crop duration, increase respiration rate, alter photosynthetic partitioning, alter phenology of flowering and fruiting and hasten the senescence, ripening and maturity. At the same time, there would be rise in new biotypes of insect and pathogen causing losses. However, through strategic research and innovations many of these problems could be effectively addressed.

I compliment the efforts of Dr H.P. Singh, DDG (Horticulture), ICAR, in addressing the issue of climate change on Indian Horticulture and developing strategies for research to address these challenges, and for compiling and editing the book entitled "***Challenges of Climate Change – Indian Horticulture***". I am sure, this publication will be of immense use to all those working for adaptation and mitigation to climate change.

(R.S. PARODA)
Chairman
Trust for Advancement of Agricultural Science
New Delhi, India

Preface

Climate change is global, but its nature, extent and magnitude are variable in different regions and locations. Hence, the issues of climate change and solution to the problems arising out of it requires local analysis, planning and management. Addressing problems of climate change is more challenging in horticulture crops compared to annual food crops. The ease with which a crop or variety can be changed or replaced for food crops is not possible in horticultural orchard and plantation, which require heavy monetary inputs and a long time to establish. Obviously a prompt response to climate change for an orchardist is very difficult. Therefore, it needed a thorough analysis and understanding about climate change at regional levels in relation to both annual and perennial horticultural crops, which could be managed through innovation, technology evaluation and refinement to provide effective solutions. We have to visualize likely changes, factors associated with change and examine the possibility to mitigate the impact of climate change through effectively planned innovative research.

Impact of climate change is being studied across the globe on water situation, ecology, wildlife and agriculture. Evidence available suggests that impact will differ from region to region, depending upon current ecological and climatic conditions. With rise in temperature, wheat productivity will decline in subtropical region, but it may increase in temperate region. Some of the evidences suggest that with rise in carbon dioxide concentration there could be an enhanced photosynthesis and ultimately higher biological yield provided water is managed effectively. However, increase in temperature may alter the photosynthate partitioning, phenology of flowering and new pathogens may emerge. Development of high temperature tolerant cultivars, change in production systems and use of new tools and technology would help in mitigating and adapting to climate change.

The potential of perennial fruit and plantation crops for carbon sequestration provide an opportunity. In the aftermath of the KYOTO PROTOCOL international treaty, providing for market based clean development mechanisms (CDM), many opportunities have emerged to earn carbon credits for mitigation of climate change.

Therefore, it is imperative that impact of climate change across the crops and regions is analyzed based on the information available, deliberated thoroughly and develop strategies to address the emerging challenges. Accordingly two days Brain Storming Session was organized to deliberate various issues at CPRI, Shimla, in September, 2008. The deliberations brought out useful information with regard to impact of climate change on horticultural crops, which is compiled in this book.

H.P. Singh
J.P. Singh
S.S. Lal

About the Editors

Born on 2nd July, 1950, Dr H.P. Singh has served horticulture in various capacities namely, Project Coordinator (Fruits); Director, NRC for Banana; Horticulture Commissioner, Govt. of India; Chairman, Coconut Development Board; Dean (Agriculture), GBPUA&T, Pantnagar; Vice–Chancellor, Rajendra Agricultural University, Pusa and now Deputy Director–General (Horticulture), ICAR.

He has also served international organisations like the Chairman of APCC, Jakarta, National Director for internationally-aided projects and Chairperson of Steering Committee of BAPNET of Bioversity International and implemented various international programmes. Besides, excellent academic record, Dr Singh has contributed to research and agricultural education which has earned him three international, twenty one national awards and five fellowships including the Fellow of NAAS. He has developed many cultivars and technologies which has shaped the horticulture. Recognising his immense contribution, OUAT, Bhubaneshwar, has conferred D.Sc. (*Honoris causa*) last year.

Dr Singh is the architect of Mission mode approach for horticultural development, starting from planning to execution which has been due to his insight. He has provided leadership to horticulture and is credited to have authored and edited 39 books and 30 bulletins, besides 175 research papers. He has widely travelled in India and abroad. His efforts have provided a strong foundation for horticultural development, leading to ***Golden Revolution*** in the country.

Born on 24th August, 1948, Dr J.P. Singh is currently Principal Scientist, at Central Potato Research Station, Jalandhar. He started his research career as SRA at Water Technology Center, IARI, New Delhi in 1972. He later joined Agricultural Research Services of ICAR in 1976 as Scientist S1 (Soil Science) at Central Potato Research Station, Patna. His main areas of expertise is crop simulation modeling, application of remote sensing and GIS for crop acreage and production estimates of potato, precision farming and potato based cropping systems. He has to his credit INFOCROP-POTATO simulation model for growth and yield of potato under short day tropical conditions. He is recipient of Indian Society of Soil Science, Zonal Award 1991; Fertilizer Association of India Silver Jubilee Award 1992; Indian Potato Association Best Paper Awards 1999, 2007; Dr. S. Ramanujam Birth Centenary Celebrations Merit Award 2004. He has published more than 150 research articles/review/book chapters/technical bulletins.

Born on 2nd July, 1952, Dr S.S. Lal is currently Principal Scientist and Head, Division of Crop Production, at Central Potato Research Institute, Shimla. He joined Agricultural Research Services of ICAR in 1976 as Scientist S1 (Agronomy) at Central Potato Research Institute, Shimla. His main areas of expertise is application of remote sensing and GIS for crop acreage and production estimates of potato, crop simulation modeling, weed management and potato based cropping systems. He is credited with organizing crop acreage and tuber yield estimation through RS and GIS. He is recipient of Potash and Phosphate Institute of Canada & FAI award, 1993; Indian Potato Association Best Paper award, 2007; Central Potato Research Institute merit award 1999, Dr. S. Ramanujam Birth Centenary Celebrations Merit Award 2004. He has published more than 175 research articles/review/book chapters/technical bulletins.

Contents

Contributors

Aishwath O.P.
National Research Centre on Seed Spices
Tabiji
Ajmer – 305206
Rajasthan

Ambast S.K.
Central Agricultural Research Institute
Port Blair – 744101
Andaman & Nicabar Islands

Anwer M.M.
National Research Centre on Seed Spices
Tabiji
Ajmer – 305206
Rajasthan

Babu Kochu M.
National Research Centre for Oil Palm
Pedavegi – 534450
Andhra Pradesh

Bai Kasturi K.V.
Central Plantation Crops Research Institute
Kasaragod – 671124
Kerala

Barman D.
National Research Centre for Orchids
Pakyong – 737106
East Sikkim

Bhargava Rakesh
Central Institute for Arid Horticulture
Bikaner – 334006
Rajasthan

Bhatt R.M.
Indian Institute of Horticultural Research
Bengaluru – 560089
Karnataka

Bhat M.G.
National Research Centre for Cashew
Puttur – 574202
Karnataka

Bhat Narayana M.
Central Potato Research Institute Campus
Modipuram
Meerut – 250110
Uttar Pradesh

Bhriguvanshi S.R.
Central Institute for Subtropical Horticulture
Lucknow – 227107
Uttar Pradesh

Chempakam B.
Indian Institute of Spices Research
Calicut – 673012
Kerala

Chowdappa P.
Indian Institute of Horticultural Research
Bengaluru – 560089
Karnataka

Das Manish
National Research Centre for Medicinal & Aromatic Plants
Anand – 387310
Gujarat

Dua V.K.
Central Potato Research Institute
Shimla – 171001
Himachal Pradesh

Govindakrishnan P.M.
Central Potato Research Institute
Shimla – 171001
Himachal Pradesh

Huchche A.D.
National Research Centre for Citrus
Nagpur – 440010
Maharashtra

Kandiannan K.
Indian Institute of Spices Research
Calicut – 673012
Kerala

Kant K.
National Research Centre on Seed Spices
Tabiji
Ajmer – 305206
Rajasthan

Krishnamurthy K.S.
Indian Institute of Spices Research
Calicut – 673012
Kerala

Kumar S. Naresh
Division of Environmental Sciences
Indian Agricultural Research Institute
New Delhi – 110012

Kumar Krishna N.K.
Indian Institute of Horticultural Research
Bengaluru – 560089
Karnataka

Lal S.S.
Central Potato Research Institute
Shimla – 171001
Himachal Pradesh

Laxman R.H.
Indian Institute of Horticultural Research
Bengaluru – 560082
Karnataka

Lawande K.E.
National Research Centre for Onion and Garlic
Rajgurunagar – 410505
Maharashtra

Mahapatro Debasish
Verve Consulting Pvt. Ltd.
Brahmeswar Bagh
Bhubaneswar – 751002
Orissa

Medhi R.P.
National Research Centre for Orchids
Pakyong – 737106
East Sikkim

Meena S.S.
National Research Centre on Seed Spices
Tabiji
Ajmer – 305206
Rajasthan

More T.A.
Central Institute for Arid Horticulture
Bikaner – 334006
Rajasthan

Mustaffa M.M.
National Research Centre for Banana
Thiruchirapalli – 620102
Tamil Nadu

Pandey S.K.
Central Potato Research Institute
Shimla – 171001
Himachal Pradesh

Panigrahi P.
National Research Centre for Citrus
Nagpur – 440010
Maharashtra

Parthasarathy Utpala
Indian Institute of Spices Research
Calicut – 673012
Kerala

Parthasarathy V.A.
Indian Institute of Spices Research
Calicut – 673012
Kerala

Rai B.
National Research Centre for Orchids
Pakyong – 737106
East Sikkim

Ravi I.
National Research Centre for Banana
Thiruchirapalli – 620102
Tamil Nadu

Raviprasad T.N.
National Research Centre for Cashew
Puttur – 574202
Karnataka

Ravi V.
Central Tuber Crops Research Institute
Thituvananthapuram – 695017
Kerala

Ravindran C.S.
Central Tuber Crops Research Institute
Thituvananthapuram – 695017
Kerala

Reddy Krishna M.
Indian Institute of Horticultural Research
Bengaluru – 560089
Karnataka

Sharma S.R.
Directorate of Mushroom Research
Solan – 173213
Himachal Pradesh

Sharma V.P.
Directorate of Mushroom Research
Solan – 173213
Himachal Pradesh

Sharma Y.K.
National Research Centre on Seed Spices
Tabiji
Ajmer – 305206
Rajasthan

Shivashankara K.S.
Indian Institute of Horticultural Research
Bengaluru – 560082
Karnataka

Shivankar V.J.
National Research Centre for Citrus
Nagpur – 440010
Maharashtra

Singh H.P.
Deputy Director General, (Hort)
Indian Council of Agricultural Research
Krishi Anusandhan Bhavan
New Delhi – 110012

Singh J.P.
Central Potato Research Station
Jalandhar – 144003
Punjab

Singh B.P.
Central Potato Research Institute Campus
Modipuram,
Meerut – 250110
Uttar Pradesh

Sridhar V.
Indian Institute of Horticultural Research
Bengaluru – 560089
Karnataka

Srinivasa Rao N.K.
Indian Institute of Horticultural Research
Bengaluru – 560089
Karnataka

Srivastava R.C.
Central Agricultural Research Institute
Port Blair – 744101
Andaman & Nicabar Islands

Suresh K.
National Research Centre for Oil Palm
Pedavegi – 534450
Andhra Pradesh

Tewari R.P.
Directorate of Mushroom Research
Solan – 173213
Himachal Pradesh

Thomas G.V.
Central Plantation Crops Research Institute
Kasaragod – 671124
Kerala

Verghese Abraham
Indian Institute of Horticultural Research
Bengaluru – 560089
Karnataka

Yadukumar N.
National Research Centre for Cashew
Puttur – 574202
Karnataka

1

Impact of Climate Change on Horticultural Crops

H.P. Singh

Global warming and climate change is now perceived to be the greatest threat to mankind in 21st century. Life and Ecosystems have evolved within a narrow band of climatic-environmental conditions in general over centuries. The natural variation in earth's climate is caused by many cosmological and geological processes. However, the 'climate change' refers to additional and relatively rapid change induced by human activities. Earth's surface temperature has risen relatively fast over the last 30 years. The rise in temperature is attributed to alarming increase in the atmospheric concentration of green house gases (GHG) CO_2, CH_4, N_2O and chloroflurocarbons mainly due to accelerated consumptions of fossil fuels in industries and transport. The average global temperature has increased by 0.8 °C in past 100 years and is projected to rise by 1.8 to 4.0 °C by the year 2100. It is now well known that warming will also be associated with changes in rainfall patterns, increased frequency of extreme events of drought, frost and flooding. Collectively all these factors will affect output of agriculture and allied sectors. India is one of the 27 countries likely to be most affected. Horticulture, in India with approximately 30% contribution in agricultural GDP from only 8% of cultivated land is threatened with serious consequences in production, quality and processing and increased cost of plant protection.

Change is a phenomenon, which is a truth, and it continues. The change could be for betterment or it could have a disastrous impact. Technological changes have provided immense comfort, and also caused the imbalances in climatic parameters, threatening the sustainability. For the last one year, various institutions, Government organizations and also the Indian Council of Agricultural Research (ICAR) have been deliberating upon the likely change in climate and its adverse impact on agriculture, and there have been clear conclusion that climate is changing, which needs public intervention and also preparedness to face the challenges.

STATUS OF HORTICULTURAL CROPS IN INDIA

Horticulture, has emerged as core sector in agriculture, passing through various phases. Investment made in horticulture during IX and X Plan have been highly productive in transforming an agrarian economy in many states, which has provided insight for reversing the trend of ever declining farmers' income and, above all addressing the nutritional security and environmental concerns. Most significant change in the last two decades have been that horticulture has moved from rural confine to commercial production, of fruits, nuts, vegetables, tubers, mushroom, spices, floriculture, medicinal and aromatic plants and plantation crops leading to use of technologies and private sector investment in production system management.

Impact of change in technologies like new cultivars and production system management is visible in terms of increased production and productivity, which has recorded seven-fold increase (185 million tonnes during 2006-07) from the level of 25 million tonnes in 1950-51. Undoubtedly, horticulture sector has moved dynamically despite numerous challenges and shortcomings, and is in crucial phase of development needing initiatives for sustainable development. To achieve the targeted production of 300 million tonnes of horticultural crops, the challenges ahead are to have sustainability and competitiveness, in the environment of declining land and water availability and threat of climate change. Vertical growth, through the use of new cultivars, efficient water and nutrient management, effective plant health management coupled with strategies for reduced post-harvest losses could be the approaches to achieve the goal. This would need appropriate innovation and investment.

Protected cultivation has shown yield enhancement up to 4 times, but would need investment. Plant architectural engineering and management can mitigate the problem associated with seasonality in many crops and will enhance efficiency of irrigation utilizing modern techniques to reduce water stress. Horticulture provides variability and has potential to adjust in different agro-climatic conditions.

What is Climate Change?

Significant variation in either mean state of climate or in its variability, persisting for an extended period (typically decades or longer) is referred as climate change, which may be due to natural internal processes or external forcing, or to persistent anthropogenic changes in the composition of the atmosphere or in land use. The earth is the only planet in our solar system that support life, because of unique environmental conditions that are present - water, an oxygen-rich atmosphere, and a suitable surface temperature. It has an atmosphere of proper depth and chemical composition. About 30% of incoming energy from the sun is reflected back to space while the rest reaches the earth, warming the air, oceans, and land, maintaining an average surface temperature of about 15 °C. The chemical concentration in atmosphere for nitrogen is 78%; about 21% is oxygen, which all animals needs to survive; and only a small (0.036%) is made up of carbon dioxide which plants require for photosynthesis. In atmosphere, energy is absorbed by the land, seas, mountains, etc., and simultaneously released in the form of infrared waves. All this released heat is not lost to space, but is partly absorbed by some gases present in very small quantities in atmosphere, called greenhouse gases (GHG) consisting of carbon dioxide, methane, nitrous oxide, water vapour, ozone and a few others. In absence of emission of heat imbalances are created. Thus, increased concentration of GHG leads to increased temperature which in turn has impact on the world climate, leading to the phenomena known as climate change.

The earth's climate system constantly adjusts so as to maintain the balance between the energy that reaches it from the sun and the energy that goes from earth back to space. This means that, even a small rise in temperature could mean accompanying changes in cloud cover and wind patterns. Some of these changes may enhance the warming, while others may counteract. Cooling effect may result from an increase in the levels of aerosols (small particles of matter or liquid that can be produced by natural or man-made activities). Positive feedback may result from an increase in water vapour due to high evaporation with rise in temperature and can further add to the warming effect. The significant change may impact agriculture/horticulture/fish/livestock and consequently food supply. Climate change per se is not necessarily harmful, but the problems arise from extreme events that are difficult to predict, like more erratic rainfall pattern and unpredicted warm spells shall affect productivity. At the same time, more availability of CO_2 would help in improved yield of root crops and increased temperature may shorten the period.

Scenario of Climate Change

Climate change is now an acknowledged fact and reality. The evidence gathered world over using state-of-the-art technology by various national and international agencies is irrefutable. Human activities like rapid industrialization, intensive agriculture, indiscriminate use of fertilizers, deforestation and increasing use of fossil fuels during past 150 years are considered major factors for climate change. The continued effect of these activities results in increasing emission of CO_2 and other greenhouse gases (GHG), leading to global warming as a 'greenhouse effect' due to entrapment of back radiation from earth by these gases. The current level of CO_2 in atmosphere, the main GHG, is 35.4% more than the pre-industrial levels and is growing. The level of greenhouse gases are now far above any level in the past 650,000 years as indicated by the study of the ice core permafrost. The concentration of all GHG reported to have increased by as much as 70% between 1970 and 2004.

The increase in temperature due to global warming is 0.76 °C since 1850. The rate of warming in the last 50 years is double than that for the last century. As many as 11 of the past 12 years were warmest since 1850, when records began. The likely increase in temperature is 1.8–4 °C by next century. The threshold value of temperature rise is 2 °C for devastating, dangerous and irreversible consequences of warming to manifest world over. Global warming is occurring along with shifting pattern of rainfall and increasing incident of extreme weather events like floods, droughts and frosting. Recent studies suggest clear evidence of reduction in light intensity, rapid melting of glaciers and rise in sea-level. The estimates of Indian Institute of Tropical Meteorology, Pune, indicates similar trends in India with slightly higher magnitude. It is projected that rainfall over India will increase by 15–40%, and the mean annual temperature will increase by 3–6 °C by the end of 21st century. Warming is likely to be more pronounced over land areas with the maximum increase over northern India. The winter and post-monsoon seasons are likely to be more affected by warming. Beside, there would be more fluctuation in temperature and water availability may decline.

Climate Change in Asia

The South Asian region is projected to be one of the most vulnerable to climate change. It is attributed to increasing population pressure, extreme poverty, predominance of agriculture and resource crunch in the region. It is projected that by 2020, food requirement in South Asia would be 50% more than the current demand. The challenge is to produce the same from constant or even shrinking land resources due to competition for land from other sectors like infrastructure, industry and housing. Poverty alleviation and food security for teeming millions in India under adverse climate change scenario would be a daunting task. These changes in global environment will have profound effects and serious consequences for agriculture, horticulture, arable eco-system and society as a whole.

Impact of Climate Change

Global climate change is expected to affect agricultural/horticultural crops through its direct and indirect effects. Scientific evidence suggests a positive effect of increase in atmospheric carbon dioxide in C3 photosynthetic pathway promoting their growth and productivity. Increased CO_2 will reduce evapo-transpiration and thus increase water-use efficiency. However, positive effects will be counteracted by increase in temperatures. Rise in temperature will reduce crop duration, increase respiration rate, alter photosynthate partitioning to economic product, alter phenology, particularly flowering, fruiting and reduce chilling unit accumulation, hasten senescence, fruit ripening and maturity. Cauliflower grows best in cool to warm conditions (15–25 °C) with high humidity. Though, some varieties can grow with temperatures over 30 °C, most varieties are very sensitive to higher temperature. Delayed curd initiation in cauliflower is reported due to increased temperature. Temperature above 30 ºC induce maximum flower and fruit drop and high temperatures after pollen release decrease fruit setting, fruit yield and seed setting in tomato.

The sensitivity of individual crop to temperature depends on inherent tolerance and growing habits. Indeterminate crops are less sensitive to periods of heat stress because the time of flowering is extended compared with determinate crops. In onion, warmer temperatures shorten the duration of growth leading to lower crop yields. Any soil warming would be advantageous for cucurbits, which are generally direct seeded and have a high heat requirement. The rise in temperature will influence survival and distribution of pest population; developing new equilibrium between alternate host crops and pests; hasten nutrient mineralization in soils; decrease fertilizer-use efficiency; and increase evapotranspiration with reduced water-use efficiency. The net effect of climate change on horticultural crops will depend on interaction effects of rise in temperature and CO_2 concentration in atmosphere. In general, CO_2 enrichment does not appear to compensate for the detrimental effects of higher temperature on yield. Most importantly, the quality of produce of these horticultural crops is likely to be impacted severely.

Many slow-growing fruit crops require heavy investment on establishment of orchards. Quick alteration/shifting of fruit species or varieties would be difficult and painful loss-bearing exercise under the impact of climate change, which may discourage the development. Recent studies have indicated that in Kullu district of Himachal Pradesh, farmers shifted from apple to vegetables, while in Shimla district at relatively higher altitude orchards have been replaced from high-chilling requiring apple cultivars of apple (Royal Delicious) to low-chilling requiring cultivars and other fruit crops like kiwi, pomegranate and vegetables. In midhills of Shimla district, trend is to shift from apple and potato cultivation totally. It is corroborated by declining trend in snowfall and apple productivity in Himachal Pradesh. The mean productivity of apple in 1980-81 was 7.06 tonnes/ha compared to 4.65 tonnes/ha in 2004-05. Since many crops with chilling requirements are tree species, moving production areas is difficult. Thus in replanting orchards and plantations over the next decade, selection of lower-chilling requiring types may be advisable. This is just an example of impending impacts of global warming and climate change.

Grape being a temperate crop has very well adapted to tropical regions, under climate change scenario with increase in temperature there would be change in growing degree days (GDD), which has direct bearing on phenology of the crop. Hence, under such circumstances, we would have to identify varieties and regions suitable for production of quality fruits. Shift in varietal choice may become necessary in case of grapevines, banana, mango and other important horticultural crops. With global warming production areas for specific crops and/or timing of planting could be changed, but for many horticultural crops, market windows and infrastructure, such as availability of local packing and distribution facilities are critical components of the production system. Locations of important production areas are often defined as much by available land, markets and infrastructure as by climatic conditions per se. Thus, as horticulturists we have to ask ourselves and our clientele whether it is realistic to move production areas in response to climate change, or whether there are other production practices that can be adjusted to compensate climate change. Climate change and CO_2 are likely to alter important interactions between horticultural plants and pollinators, insect and disease, and pests and weeds.

We need quick and clear understanding of impact of climate change on horticultural crops for making sound action plan because horticulture based farming systems have high potential for sequestering carbon for mitigation of climate change. The perennial trees act as carbon sinks by sequestering the atmospheric carbon. The carbon credits could be earned under the clean development mechanism (CDM). The horticultural waste could be composted locally instead of dumping in the landfills, which can reduce the release of global warming methane that is involved in global warming. The organic waste could also be used for generating biogas as an alternate energy source. There are considerable uncertainties about agronomic implications of horticultural crops. Predicting impact of climate change on horticultural crops accurately on regional scale is a big problem. It can be accomplished only by a modeling approach through well-validated robust crop simulation models. These crop simulation models incorporate the effect of various factors of growth and yield in a mathematical model processed by computers to give results quickly for specific situation. Well-validated simulation tools developed for cereal crops have been helpful in predicting of impact of climate change.

Modeling for Better Understanding of Impact

Availability and development of good simulation models for horticultural crops (fruit and vegetables) is lacking in India probably with exception of potato and coconut. In tomato and onion crops, the InfoCrop model has been adapted and the model is being validated for different agro-ecological regions. Perennial nature of large-sized fruit trees and shrubs are problematic in study of direct effect of various factors of growth, development and yield in controlled environment. Innovative methods are required to develop simulation models for important horticultural crops like mango, grape, apple, orange, citrus, litchi, guava, etc., on priority. Similarly, annual vegetable crops like capsicum, cucurbits, cauliflower and root and tubers, and flower crops urgently need crop simulation models. Once these simulation models are available, prediction of vulnerability of existing areas under these horticultural crops to climate change scenario can be examined and new target areas for possible shifting of species and varieties for cultivation can be identified. Possible adaptation measures to reduce the impact of climate change are also possible to study through simulation models for suggesting changes in management practices. Development of crop simulation models for horticultural crops in India is now a priority area of research.

Adaptation Strategies

To address the adverse impacts of climate change on productivity and quality of horticultural crops we need to develop sound adaptation strategies. The emphasis should be on development of production systems for improved water-use efficiency and to adapt to the hot and dry conditions. Strategies like changing sowing or planting dates in order to combat the likely increase in temperature and water stress periods during the crop-growing season. Modifying fertilizer application to enhance nutrient availability and use of soil amendments to improve soil fertility and enhance nutrient uptake. Providing irrigation during critical stages of the crop growth and conservation of soil moisture reserves are the most important interventions. The crop management practices like mulching with crop residues and plastic mulches help in conserving soil moisture. In some instances excessive soil moisture due to heavy rain becomes major problem and it could be overcome by growing crops on raised beds.

In addition to employing modified crop management practices, the challenges posed by climate change could be tackled by developing tolerant varieties. Several institutions have evolved hybrids and varieties, which are tolerant to heat and drought stress conditions. They must be used very effectively to combat the effect of climate change depending upon their performance in a given agro-ecological region. Efforts should be intensified to develop new varieties suitable to different agro-ecological regions under changing climatic conditions. In comparison to annual crops, where the adaptation strategies can be realized relatively fast using a wide range of cultivars and species, changing planting dates or season, the planting and rearrangement of orchards requires a consideration of the more long-term aspects of climate change. Therefore, before resorting to any adaptation option, a detailed investigation on the impact of climate change on perennial crops is necessary.

The long-time horizon of perennial crops creates situations like; favourable areas may become unfavourable during the life of a single orchard. The choice of a variety is complicated by the risk that the best variety for the current climate may be poorly suited for future climates. Thus, while adaptations such as planting new varieties and shifting to new areas may reduce impacts in the long-term, short-term losses may largely be unavoidable. In wine grape, each grape variety grows in a range of temperatures and for each variety it is possible to define climates for premium wine production. The physiological and morphological differences between varieties (genotypes) enable production over a relatively large range of climates and depending upon the suitability to different growing areas the cultivars may be adopted. In situations, where there is a strong consumer preference for a select cultivar and also the suitable varieties are not available to adapt to the changing climate of a particular growing region, the option of using rootstocks for better performance of the scion cultivars could be explored.

Mitigating Climate Change — the Past Experience

Although climate change is a reality, CO_2 and methane are likely to increase which may cause impact in terms of increased temperature, more demand for water and increase biotic and abiotic stresses. On midhill chilling will not be enough to induce flowering in apple and high temperature may cause desiccation in pollen, shriveling of fruits resulting in reduced yield and more failure of the crops. These are the likely impact which causes the concerns. But, there are innumerable examples to cite that, climate has been changing and the technologies have helped in mitigating the problem. In matter of fact, grape is a temperate fruit, which has been largely grown under cool climate, be it for table purposes or for wine-making. But the technological change in plant architecture and production system management has helped to produce grape in tropical situation, with highest productivity in the world. Salinity and alkalinity were a great problem for successful growing of grape but identification of suitable rootstocks has made it highly productive.

If we look to potato, tomato, cauliflower and cabbage, these are thermo-sensitive crops and were productive only under long day conditions in temperate climate. But development of heat tolerant cultivars and adjustment in production system management has made it possible with very high productivity, even in subtropical and mild subtropical and warmer climates. These are the past experiences, which clearly brings home the point that through innovative research threat of climate change could be converted into the opportunity, but will need visualization of likely change, its impact and planning to mitigate it bad impact. Now, available tools of biotechnology could add for speedier delivery of research results.

FUTURE STRATEGIES

Depending on the vulnerability of individual crop and the agro-ecological region, the crop-based adaptation strategies need to be developed, integrating all available options to sustain the productivity. Developing strategies and tools to comprehensively understand the impact of climate change and evolve possible adaptation measures in horticultural crops is less understood. To enhance our preparedness for climate change and to formulate a sound action plan, we need to identify gaps in vital information, prioritize research issues from point of view of farmers, policy-planners, scientists, trade and industry.

It is imperative to deliberate upon the likely changes which can happen in next 50–100 years, how these changes could affect growth, development and quality of horticultural crops, what are the technologies which shall help to mitigate the problem and what kind of innovative research should be done to overcome the challenges of climate change. Thus, policy issues, adaptation strategies and mitigation technologies could be worked out and challenges could be converted into opportunity with the updation of following:

- Priority of education, research and development, and policy implications for enhancing adaptive capacity of Indian horticulture to climate change.
- Capacity assessment of Indian horticulture for mitigation of GHG.
- Appropriate short and long-term action plan to mitigate the impact of climate change in horticulture.

OUTLOOK

Climate change per se will have impact on horticultural crops, due to erratic rainfall, more demand for water, and enhanced biotic and a biotic stresses. However, the changes will not be only harmful, as enhanced CO_2 concentration may enhance faster photosynthesis and increased temperature may hasten the process of maturity. However, measures to adapt to these climate change-induced changes are critical for sustainable production. Increased temperature will have more effect on reproductive biology and reduced water may affect the productivity but adaptive mechanism like time adjustment and productive use of water shall reduce the negative impact. The strategies must have to identify the gene tolerant to high temperature, flooding and drought, nutrient efficient cultivars and production system for efficient use of nutrient water. Strategies have to address the enhanced water-use efficiency, cultural practices that conserve water and promote crop. Development of climate resilient horticultural crops which are tolerant to high temperature, moisture stress, salinity and climate proofing through genomics and biotechnology would be essentially required. This would need highly prioritized research to address the impact of climate change.

Enhancing the adaptation of tropical production system to changing climatic condition is a great challenge and would require integrated efforts and an efficient and effective strategy to be able to deliver technologies that can mitigate the effects of climate change on diverse crops and production systems. Knowledge of carbon sequestration especially through perennial horticulture needs to be enhanced, which could be utilized for enhancing the income through trading of carbon. Research must address all the strategies which can convert the challenges into the opportunity. Concerted and integrated efforts with effectiveness and efficiency would definitely make us stronger to face the challenges and meet the ever increasing demand for food.

2

Climatic Change and Plantation Crops: Impact, Adaptation and Mitigation with Special Reference to Coconut

S. Naresh Kumar, K. V. Kasturi Bai and G. V. Thomas

PLANTATION crops, being perennial in nature, have to face the impact of climate change even during a single generation or in a standing plantation. Hence, it is important that the impact of climate change on plantation crops is understood well. For example, a seedling of coconut, with all likelihood, may face the increased CO_2 concentrations, temperatures, etc., in next 50 years of its economic yield producing life span. Apart from this, plantation crops mainly coconut, rubber, tea, coffee, cashew, oil palm, etc., are grown in ecologically sensitive areas such as coastal belts, hilly areas and areas with high rainfall and high humidity. All these crops are of high economic value, contributing substantially to agricultural exports at global level and at national level. Further, spice crops, generally grown as the intercrops in these plantation crops are also influenced by climate change. In India, plantation crops, viz., coconut, cashew, rubber tea, coffee, arecanut, cocoa, oil palm, etc., and spice crops like black pepper, cardamom, clove, etc., are grown in large areas of Kerala, Karnataka, Tamil Nadu, Andhra Pradesh, Maharastra, West Bengal and Assam. These plantations provide sustenance to millions of farmers across these states (Table 2.1).

Table 2.1. Area under plantation crops in India

Crop	*Area*	*Production*	*Productivity*	*Major states*
Coconut	19.46 lakh ha	14,811.1 million nuts	7608 nuts/ha/year	Karnataka,Kerala,Tamil Nadu and Andhra Pradesh
Arecanut	3.65 lakh ha	4.3 lakh tonnes	1148 kg/ha	Karnataka and Kerala
Cocoa	15,000 ha	8,000 tonnes of beans	5.3 q/ha	Karnataka and Kerala

Source: Department of Agricultural Statistics and Ministry of Commerce

IMPACT OF CLIMATE CHANGE ON PLANTATION CROPS

Climate change impact assessment in plantation crops is a big challenge as simulation models for many plantation crops are not available and even if available, they are not suitable for climate change studies. Even though simulation model in perennial crop like cocoa (Zuidema *et al.*, 2003) has been reported, the model is not yet validated for Indian conditions and suitability for climate change studies. Only recently simulation model for coconut is developed (Naresh Kumar *et al.*, 2009a), which is being used for climate change studies as well. Efforts are being made to develop the regression models for studying the climate change impacts on the productivity of arecanut and cocoa.

Quantification of impact of major climate change parameters such as increased temperature and CO_2 on crops is being studied by two approaches. In field experimentations, quantifications are done by growing crops in Open Top Chamber (OTC) or in Free Air Carbon Dioxide Enrichment (FACE) and Free Atmospheric Temperature Elevation (FATE) facilities. The other approach is to use well validated simulation models. Simulation models are strong tools which provide opportunity to use various climate change scenarios in combination with different management parameters for analyzing the regional impacts. More so in case of plantation crops as field experimentations mentioned above are not only expensive but also are not tangible.

Models in Plantation Crops

Coconut: Forecasting coconut yield is of vital importance for policy-makers, scientists, and growers, and models based on climatic variables are more meaningful than those developed without them. Peiris (1991) developed a model to predict yield a year ahead based on six variables derived from monthly rainfall distribution. A predictive model with six climatic variables was developed by Vijaya Kumar *et al.* (1988). Naresh Kumar *et al.* (2009a) developed weather data based yield forecasting models with R^2 above 0.9 for different agro-climatic zones of coconut-growing areas and their validity was tested successfully for two years. Obtained yield levels were within the range forecast using 1 to 4 year lag periods in all cases. Recently, for the first time, coconut simulation model is developed by Naresh Kumar *et al.* (2009b), which is being used for climate change studies.

Tea: Based on the fuzzy comprehensive judgment to analyse the ground climatic data in China, the areas in province of Yunnan were identified for tea cultivation (Yang *et al.,* 1993). The simulation of winter bud dormancy was done (Omae and Takeda, 2003). Model simulations of potential yield of tea in north-east India predicted mean reduction in yield up to 1.5 kg/ha for each 1 mm reduction in evapotranspiration (Panda *et al.*, 2003). CUPPA tea model is developed for modelling the seasonal yield variations (Matthews and Stephens, 1998).

Coffee: In Brazil, using geographical information system capabilities and a water balance model, the integration of climatological data allowed the identification of regions suitable for coffee cultivation (Assad *et al.*, 2000). A model was developed to reliably monitor and assess drought impact on coffee yields just before the beginning of the maturation growth stage. (Picini *et al.*, 1999).

Rubber: Modelling of rubber growth as a function of climate and soils was done based on the GROWEST classification system. Rubber yield prediction model RRYIELD, a rubber based agro-forestry model involving bioclimatic, topographic and silvicultural variables can incorporate latex and wood production and rice intercropping (Thomas *et al.*, 1995). Application of fuzzy logic to land suitability for rubber production in peninsular Thailand (Van Ranst *et al.*, 1996).

Cocoa: FST based simulation model is developed for cocoa by Zuidema *et al.* (2003). However, this model needs to be calibrated and validated for Indian conditions and suitability of this model for climate change studies is yet to be tested.

IMPACT ANALYSIS

Elevated CO_2 and Temperature

Productivity

Growth response of coconut seedlings: Among 3 cultivars (WCT, LCT and COD) and 2 hybrids (WCT × COD, COD × WCT), COD has shown early response to elevated CO_2 (500 and 700 ppm) and elevated temperature (2 °C) after 3 months of exposure. Increase in shoot height, leaf area and soot dry matter due to elevated CO_2 was to the tune of 36% over chamber control. Decrease in above parameters in elevated temperature (2 °C) was to the tune of 40% over chamber control (Naresh Kumar, 2008).

Elevated temperature (2 °C)

Open field (control)

Elevated CO_2 (500 ppm)

Elevated CO_2 (700 ppm)

Coconut seedlings in OTCs

Simulation studies for impact on yield: The climate change studies on plantation crops are scarce world over and only recently, research initiatives have been made. The current knowledge available regarding climate change studies on coconut in India are as follows.

Simulation studies to analyze the influence of increased atmospheric CO_2 at 550 ppm along with increase in temperature by 1 °C indicated increase in coconut yields by around 5–12% depending upon the agro-climatic zone and management (Table 2.2). However, these estimates are over emphasizing the carbon fertilization benefits due to large increase in CO_2. However, if CO_2 is to increase to this level, then temperatures are likely to increase beyond 1 °C causing reduction in yields in many locations. Another shortcoming in this analysis is the rainfall in future is assumed to be similar as of past 30 years. Thus, the impact analysis was done using global climate model (GCM) scenario outputs, for better projections of the future coconut yields as given in the following sections.

Table 2.2. Influence of temperature (+1 °C) and CO_2 (550 ppm) on coconut yield in India

State	*Possible change in yield during 2002-03* (%)
Kerala	8-11
Karnataka	6-11
Tamil Nadu	5-12
Andhra Pradesh	6-10
Maharastra	7-9
Gujarat	7-8
Orissa	7-10
North East	7-10

FUTURE CLIMATE SCENARIOS FOR COCONUT AREAS IN INDIA BASED ON HADCM3-PRECIS MODEL OUTPUTS

Using the historical weather data as baseline (1960-1990), the climatic scenarios for different coconut-growing areas were analyzed based on the HadCM3 scenarios. For example, in PRECIS –A2 scenario for Kerala, for 2071 to 2100 indicate increase in day temperature by 3-4 °C. Up to 1990, the Tmax reached around 37 while in 2071 to 2100 period the Tmax is projected to be around 41 °C. Similarly, Tmin, which was below 30 °C is projected to increase beyond 32 °C. Generally, the west coast of India is characterized with gentle wind velocities. However, it is projected that the wind velocities will be more with frequent winds of above 5 m/sec wind speed. This will have implications on RH leading to more drier days (less RH). The regional climate mode (RCM) based projections also indicate the occurrence of drier days. However, the total rainfall is projected to remain more or less same. Similar analysis for coastal Andhra Pradesh indicated increase in Tmax over baseline period and also increase in wind velocity.

Analysis indicated maidan parts of Karnataka, Eastern Tamil Nadu, lower coastal Andhra Pradesh, upper coastal Andhra Pradesh, Pondicherry, West Bengal and Assam in decreasing order are found to be hot spots as per different story lines (A2a, A2b, A2c, A1f, B2a, B2b) of climate change (HadCM3 model) for 2020, 2050 and 2080 scenarios.

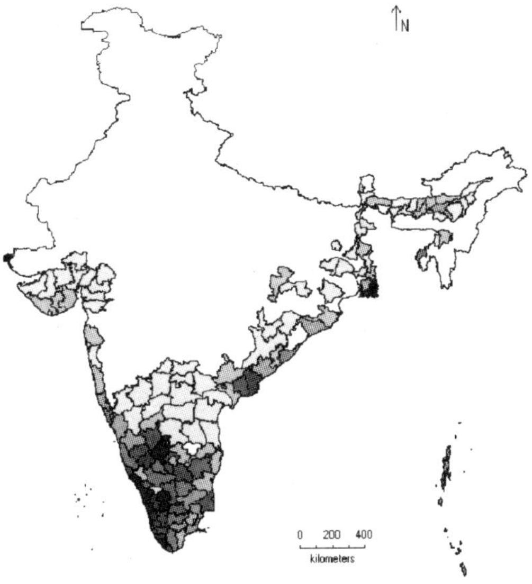

Fig. 2.1. Coconut-growing areas in India

Impact of Climate Change on Coconut Yield

Using coconut simulation model, yields were simulated for 13 agro-climatic zones represented by 16 centres (Naresh Kumar and Aggarwal, 2008a). These areas contribute over 90% to the coconut production in India. Simulations for yield projections were done for HadCM3 climate change storylines, viz., A2a, B2a and A1F for 2020, 2050 and 2080 scenarios (Fig. 2.2). The CO_2 projections of Bern CC model, temperature and rainfall projections of HadCM3 model, location weather data for past 30 years, major soil type of the agro-climatic zone and currently followed farmers' practice for crop management are used as inputs into the coconut simulation model. Relative impacts on yield were worked out at district level in a agro-climatic zones and up-scaled to state and national projections for impacts of climate change on coconut yield. Sensitivity analysis for CO_2 and temperature was also done using simulation model.

Results indicate that under all storylines, coconut productivity is projected to go up by up to 10% during 2020, up to 16% in 2050 and up to 36% in 2080 over current yields only due to climate change. However, in east coast yield is projected to decline by about 2% in 2020, 8% in 2050 and 31% in 2080 scenario over current yields due to climate change. Special variations exist for these projections. Yields are projected to go up in Kerala, Maharashtra and parts of Tamil Nadu, and Karnataka, Maharashtra while they are projected to decline in Andhra Pradesh, Orissa and Gujarat (Naresh Kumar and Aggarwal, 2008a). These projections assume that current management and irrigation will be made available in future scenarios as well. However, in the event of reduced availability of irrigation sources in future, these projections in increase in yields may be altered particularly in presentably irrigated areas.

It is also estimated based on the past 10-year domestic consumption trend that coconut consumption per capita may decline. However, projections show that demand for coconut may

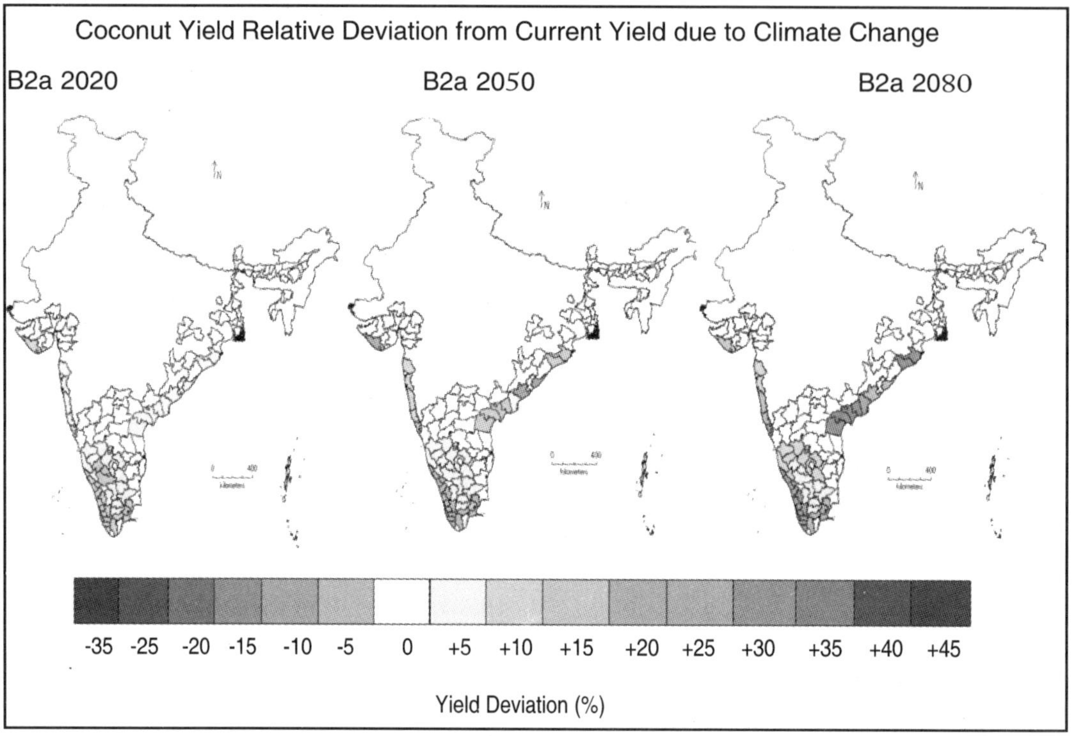

Fig. 2.2. Coconut simulation model derived relative deviations in nut yields due to climate change in B2a scenario of HadCM3 model.

increase due to increase in population in the country. This shows that demand supply gap will increase in future even as coconut production is expected to rise due to climate change alone with existing management being practised by farmers. Thus to improve the productivity, more adoption of farming technologies by farmers is essential, not only to bridge the supply demand gap but also to exploit the export opportunities.

Recent Trends in Weather, Climate, Production and Productivity of Coconut

The analysis of past weather data from different locations representing the major coconut-growing agro-climatic zones and yield data from the respective districts (Table 2.3) indicated warming trends in most of the coconut-growing areas (Naresh Kumar *et al.*, 2009b). The increase in average maximum temperature varied from 0.01–0.13 °C/year. On the other hand, average minimum temperatures are decreasing in many places. The range in change varied from -0.03–0.03/year. Dry spells are in increasing trends in Karnataka and Kerala, whereas it is reducing trend in coastal Andhra Pradesh and coastal Maharashtra. Change in dry spell varied from -1.98–0.27 days/year. Coconut productivity increased over past 50 years except recent declining trends in maidan Karnataka and Coimbatore district (Tamil Nadu) due to consecutive droughts. Change in coconut yields across the country ranged from -114–270 nuts/ha/year.

Table 2.3. Past trends in weather and productivity in coconut-growing areas in India

Centre	*Productivity*	*Temperature Max.*	*T Min.*	*>33 °C*	*< 15 °C days*	*RH days*		*(Rainfall) Annual*	*intensity <2.5 mm*	*Overall Days*	*Dry change*	*Coconut spell*
Ratnagiri (Maharashtra)	↑	↓	↑	↑	nil	↓ (RH Max)	↑	↓	↑	Cool Night Warm DayDry	↓	↑
Mulde (Maharashtra)	↑	↑	↑	↑	↓	↑	↓	↓	↓	Warming	↓	↑
Arsikeri (Karnataka)	↑	↓	↑	↑	↑	↓(RH max) Warm DayDry	↓	↑	↓	Cool Night	↔	↓
Kidu (Karnataka)	↑	↑	↓	↑	↓	NA	↓	↑	↑	Warming	↑	↓
Kasaragod (Kerala)	↑	↓	↑	↑	↑	↓	↓	↓	↑	Cool Night Warm Day	↑	↑
Trissur (Kerala)	↓	↔	↓	↓	nil	↔	↓	↑	↓	RF Distribution changing	↓	↑
Trivendrum (Kerala)	↑	↑	↑	↑	nil	NA		↓	↓	Warming	↑	↑
Aliyarnagar (Tamil Nadu)	↑	↑	↑	↑	↑	↑ RH min	↓	↓	↑	Cool Night Warm Day	↓	↓
Veppankulam (Tamil Nadu)	↑	↑	↑	↑	nil	↓ RH min	↓	↑	↓	Warming	↑	↑
Ambajipeta (Andhra Pradesh)	↑	↑	↓	↑	↔	↑ Min RH	↓	↓	↑	Warming	↓	↑
Calcutta (West Bengal)	↓	↓	↓	↓	↓	NA	↓	↓	↑	Cooling	↓	↑

↑ Indicates increase; ↓ indicates decrease; ↔ indicates no significant change

Quality: The copra yield/nut, oil percentage and fatty acid profiles of various cultivars grown at different agro-climatic zones were analyzed for their relationship with the weather during nut growth periods for the nuts harvested during January, April, July and October. The normalized deviations of these parameters were Z plotted against the normalized deviations of weather parameters during the copra development period (Fig.2.3). The analysis indicated impact of Tmax, Tmin, Rainfall, RH on copra yield/nut, oil % and fatty acid composition in coconut oil. The normalized deviations did not show any significance among HadCM3 scenarios and current trends.

Copra drying temperatures did not influence the fatty acid profile in coconut oil except of some long chain fatty acids which are in very low concentration. Variation in MCFAs in WCT grown at different agro–climatic zones was found maximum in nuts harvested during January, October and July and least in those harvested in April. Among the agroclimatic zones, the seasonal variation for LCSFAs was maximum in Ambajipeta and minimum at Kasaragod and Ratnagiri (Naresh Kumar, 2006).

Pests of target crop: Information on the impact of climate change on pests of coconut is not readily available and scarce.

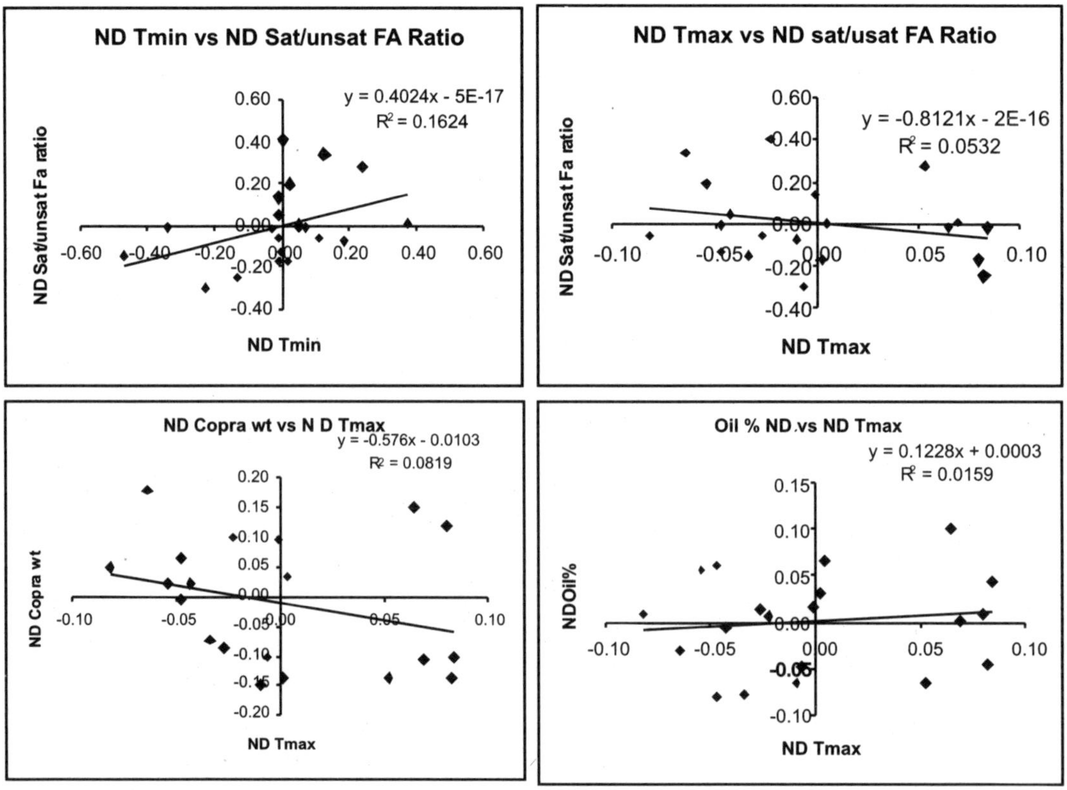

Fig. 2.3. Trend analysis of coconut quality parameters and weather

Aberrant Weather and Extreme Eevents

Droughts and cyclone: Analysis of data to study the impact of climate change related events like cyclones and droughts on coconut indicated that consecutive drought years in Coimbatore district of Tamil Nadu and Tumkur district of Karnataka caused not only drastic reduction in yield but also left about 2 lakh of palms dead. More than 6 lakh palms were severely affected. Consecutive droughts in Coimbatore district reduced the coconut production by about 3 lakh nuts/year for four years (Table 2.4). Productivity loss was to the tune of about 3,500 nuts/ha/year (Naresh Kumar *et al.*, 2007b). Length and frequency of dry spell has negative impact of coconut yields in different agro-climatic zones (Naresh Kumar *et al.*, 2007a)

Table 2.4. Coconut production loss in Coimbatore district due to consecutive droughts

Year	*Loss in production due to drought (lakh nuts)*
1999-00	-2145
2000-01	-2129
2001-02	-2661
2002-03	-5379
Total loss (lakh nuts in 4 years)	-12313
Loss (lakh nuts)/year	-3078

In Godavari districts of Andhra Pradesh, cyclone in 1995 badly affected the coconut yields. The yields drastically reduced in 1995-96 and then it took 4-5 years for the recovery of affected gardens. Reduction in coconut yields was to the tune of 3,350 lakh nuts/year for 6 years. The loss in East Godavari district was to the tune of about 2,200 lakh nuts/year for six years. The productivity was reduced by 6,200 nuts/ha/year in East Godavari district and by -4,100 nuts/ha/year in Andhra Pradesh. During 2003-04, only the production levels could reach pre-cyclone period thus severely affecting the production for six years (Table 2.5).

Table 2.5. Coconut production loss in Andhra Pradesh due to 1996 cyclone

Year	*Srikakulam*	*E. Godavari*	*W. Godavari*	*Andhra Pradesh*
1996-97	-28.9	-4245.0	-1203.7	-5753.4
1997-98	-456.3	-3400.0	-1089.2	-4943.8
1998-99	-604.5	-3313.4	-1332.0	-5726.3
1999-00	-706.9	-1289.2	17.9	-2019.1
2000-01	-406.0	-963.6		-1232.4
2001-02	-71.4	-208.0		-394.3
Loss in 6 years lakh nuts	-2274.17	-13419.3	-3607.06	-20069.5
Loss lakh nuts/year	-379.0	-2236.5	-901.7	-3344.9

ADAPTATION MEASURES FOR CLIMATE CHANGE AND GLOBAL WARMING

Soil Water Management Practices

Since climate change is projected to increase the frequency of extreme events like droughts, heavy rainfall events, etc., it is important to have proper soil and water conservation practices in place to minimize the adverse effects of climate change. Soil moisture conservation can be done at various scales. These are

- *Water management at national level*: This includes construction of irrigation dams, linking of rivers where ever ecologically feasible and creating irrigation canal system.
- *Water management at river basin level*: This includes providing proper water storage and distribution canal system with minimal loss of water during distribution process.
- *Water shed management*: This includes construction of check dams mainly for improving the ground water-table and for soil conservation.
- *Field level*: Contour bunding in undulating topography, run-off harvesting, storage and recycling, growing of cover crops help in reducing soil and water loss.
- *Farm level:* Soil and moisture conservation/management practices include drip irrigation, sub-surface irrigation, mulching of field/inter-plant spaces. Three or four drippers can be placed in opening of bamboo/tube stacked into soil at 20–30 cm depth in the palm basin so as to wet sub-soil layer. Drip irrigation is shown to increase not only field WUE but also the physiological WUE of plant (Naresh Kumar *et al.,* 2002). If late rains occur, pulses or fodder crops can be sown in between coconut rows. After harvest, these plant residues can be used as mulches. Similarly, green manure crops can be raised in the plantations. Ploughing back tender *Glyrecedia* has given encouraging results.
- *Plant level*: In plantation crops, soil and moisture conservation at plant level becomes important as well. This can be done by mulching the plant basins with leaves, farm waste, burying the coconut husks in 3 or 4 layers or composted coir pith (50 kg/palm) (Naresh Kumar, 2004). These conservation practices led to improved nut yields (Naresh Kumar *et al*., 2006). Application of green manures and organic manures (FYM) – 20 tonns/ha (50–100 kg/palm) or addition of tank silt at 100-200 kg/palm improves soil organic matter and water-holding capacity.

  Apart from this, growing water use efficient and heat tolerant cultivars is the most important measure. It is important to identify and conserve the ecotypes, which could withstand the vagaries of nature and still performed better in terms of growth and yield. In coconut-growing areas, in situ drought tolerant palms were identified and characterized (Naresh Kumar *et al*., 2004). Such palms are having desirable canopy shape, leaf number apart from being better yielder. The physiological water-use efficiency of these palms is also high. This type of in situ tolerant plants with desirable traits should be used in breeding for drought tolerance. The plants which can withstand the natural occurrence of drought and other stresses and still produce good yield are of premium in nature for their possessing desirable gene pool.

  Based on several physiological, biochemical, anatomical and morphological characters, cultivars like WCT, LCT and hybrids WCT × COD, LCT × GBGD, LCT × COD were identified as drought tolerant types.
- At sub-pant level, the management includes, cutting off the senescent lower fronds in coconut to reduce the water loss through transpiration.

- If adequate water is available, irrigation with 200 litres water/palm once in 4 days and mulching with dry leaves should be followed. But flooding the basins should be avoided. If water resources are good, saving water for facing prolonged drought should be emphasized.

The above practices which lead to soil-water-plant management, soil and water conservation and to minimize the adverse effects of climate change related events.

Planting in Recommended Depth and Spacing

- In areas prone to cyclones and strong winds, growing of *Casurina* in several rows helps as wind break and reduce the intensity of cyclone.
- To minimize the loss due to lodging and uprooting of palms during cyclone/strong wind events, planting of seedlings in 1 m^3 pit at recommended depth is desirable.

Climate Related Management

- Growing coconut on raised bunds to overcome adverse effects due to water-logging in low-lying areas.
- Covering arecanut stem with leaves to minimize the damage due to intensive solar radiation.

Identification and Development of 'Adverse Climate Tolerant Cultivars'

Adaption of crop intensification, crop diversification and integrated farming approach for combating the climate change risks.

Mitigation Measures to Reduce Emission of CO_2/GHG and Corbon Sequestration Potential of Crops

- *Identification of eco-friendly and green low cost technologies*: It is essential that the new agro-technologies for crop management should be ones with minimal GHG emissions. Soil moisture conservation measures, slow release fertilizers, minimal tillage, recycling of biomass, enrichment of soil organic carbon should be given priority.
- Efficient and minimal utilization of fuel for on-farm and farm related operations, post-harvest transportation, post-harvest processing and consumption stages.
- Efficient conversion and recycling of farm biomass and stop burning the farm biomass.

Carbon Sequestration and Stock in Plantation Crops

Plantation crops offer to be good candidates for carbon sequestration and carbon trade. However, currently, under Clean Development Mechanism (CDM), plantation crops are not included for carbon trade. Keeping future possibilities on inclusion of plantations for carbon trade, carbon sequestration potential of plantation crops such as coconut, arecanut and cocoa was worked out.

The whole plant carbon stocks and sequestrations were analyzed and worked out for coconut plantations under different agro-climatic zones (Naresh Kumar, 2007, 2008a and b). Apart from these, variations due to cultivar, agro–climatic zone and management were also analyzed. The results indicate that annual carbon sequestration in coconut above ground biomass varied from 15–35 tonnes CO_2/ha/year depending on cultivar, agro-climatic zone, soil type and management. Annually sequestered carbon stocked into stem in the range of 0.3–2.3 tonnes CO_2/ha/year. Standing carbon stocks in 16 year-old-coconut cultivars in different agro-climatic zones varied from 15 to 60 tonnes CO_2/ha/year. Annual carbon sequestration by coconut plantation is higher in red sandy loam soils and lowest in littoral sandy soils (Naresh Kumar, 2007, 2008a and b). Cocoa-arecanut also is a good system for carbon sequestration with a potential to sequester 5 to 7 tonnes CO_2/ha/year (Balasimha and Naresh Kumar, 2009).

Intensification of plantations with other perennial crops will help in sequestering more carbon. This should be synergized with the adaptation of green technologies in every step of production to consumption chain.

FUTURE STRATEGIES FOR RESEARCH

- Elaborate studies on impact of elevated CO_2 and temperature on coconut, arecanut and cocoa seedlings/adult palms using OTC facility.
- Assessing the impacts, adaptation and vulnerability of coconut to different climate change scenarios using coconut simulation model.
- Estimation the carbon sequestration and stock potential of coconut plantations in different agroclimatic zones using simulation model.
- Carbon trade mechanism for coconut, arecanut and cocoa plantations.
- Identification and development of eco-friendly technologies for reduced GHG emissions during entire chain from production to consumption.

CONCLUSION

Research efforts indicated that the climate is changing in areas under plantation crops. Past adverse climate events such as droughts and cyclones severely affected the coconut yields. Coconut simulation model based projections as per HadCM3 scenarios indicate beneficial influence of climate change on coconut in western coast of India provided the current water sources are made available in future as well. On the other hand, the plantations in irrigated areas of Tamil Nadu, Karnataka, Andhra Pradesh, Orissa and Gujarat are likely to be adversely affected causing loss of production. Seedlings of coconut, arecanut and cocoa have shown positive response to elevated CO_2 in open top chamber (OTC) experiments. Quality of the produce is also likely to be affected with changes in nut composition and fatty acid profile of oil. Soil moisture conservation needs to be given paramount importance for crop production in water limiting scenarios of future. Apart from this, development of '*green technologies*', '*adverse climate tolerant varieties*', '*adverse climate proofing*' by crop diversification, crop intensification, mixed farming need to be promoted. The carbon sequestration potential of plantation crops needs to be exploited for acquiring carbon credits for the overall benefit of farmers.

References

Assad, E.D., Evangelista, B.A., Silva, FAM. da, Santos, E.R. dos, da-Silva FAM, dos-Santos E.R. 2000. Climatic zoning of coffee (*Coffea arabica*) for Goias State. *Comunicado Tecnico Embrapa Cerrados.* 2000, No.35, 6 pp.

Balasimha, D. and Naresh Kumar, S. 2009. Net primary productivity, carbon sequestration and carbon stocks in areca-cocoa mixed crop system. *Cocoa conference.*

Matthews, R.B. and Stephens, W. 1998. CUPPA -TEA: A simulation model describing seasonal yield variation and potential production of tea. 2. Biomass production and water use. *Experimental Agriculture* **34**:369-89.

Naresh Kumar, S., Rajagopal, V., Laxman, R.H., Dhanapal, R. and Maheswarappa, H.P. 2002. Photosynthetic characteristics and water relations in coconut palms under drip irrigation levels. *In: 'Plantation Crops Research and development in new Millennium'* Rathinam, P., Khan,H.H., Reddy, V.M., Mandal, P.K. and Suresh, K. (Eds). Proceedings of PLACROSYM XIV Coconut Development Board Pub., pp116-20.

Naresh Kumar, S., Rajagopal, V., Siju Thomas, T. and Cherian Vinu, K. 2006. Effect of conserved soil moisture on the source-sink relationship in coconut (Cocos nucifera L.) under different agro-climatic conditions in India. *The Ind. J. Agric. Sci.* **76:**277-81.

Naresh Kumar, S., Rajagopal, V., Siju Thomas, T., K. Cherian Vinu, Ratheesh Narayanan, M.K., Ananda, K.S., Nagawekar, D.D., Hanumanthappa, M., Vincent, S. and Srinivasulu, B. 2007a. Variations in nut yield of coconut (*Cocos nucifera* L.) and dry spell in different agroclimatic zones of India. *Ind. J. Hort.* **64**:309-13.

Naresh Kumar, S. Sairam, C.V., Jose, C.T., Kasturi Bai, K.V., Palnaiswamy, C., Rajagopal, V., Krishnamurthy, K.S., Kandiannan, K. and Champakam, B. 2007b. Climate change effects on growth and productivity of plantation crops with special reference to coconut and black pepper: Impact, adaptation and vulnerability and mitigation strategies- ICAR Network Project on Impact of Climate Change on Indian Agriculture entitled Final Report pp. 21.

Naresh Kumar, S., Kasturi Bai, K.V., Rajagopal, V. and Aggarwal, P.K. 2008. Simulating coconut growth, development and yield using InfoCrop-coconut model. *Tree Physiology* (Canada) **28**:1049-58

Naresh Kumar, S., Rajagopal, V., K. Cherian Vinu, Siju Thomas, T., Sreenivasulu, B., Nagvekar, D.D., Hanumanthappa, M., Bhaskaran, R., Vijaya Kumar, K., Rateesh Narayan, M.K. and Amarnath, C.H. 2009a. Weather data based descriptive models for prediction of coconut yield in different agro-climatic zones of India. *Indian J Hort.* (in press).

Naresh Kumar, S., Rajeev, M.S., Vinayan Nagvekar, D.D., Venkitaswamy, R., Raghava Rao, D.V., Boraiah, B., Gawankar, M.S., Dhanapal, R., Patil, D.V. and Kasturi Bai, K.V. 2009b. Trends in weather and yield changes in past in coconut growing areas *Indian. J. Agromet.* (in press)

Naresh Kumar, S. 2004. Drought management in coconut gardens. (English)– CPCRI publication–16pp.

Naresh Kumar S. 2007. Carbon sequestration potential of coconut plantations. *ICAR NEWS.* Jul-Sept. **13**:15-16.

Naresh Kumar, S. 2008a. Modeling impacts and adaptations in coconut to climate change in different agro-climatic zones and field response of coconut to increased temperatures and CO_2. Project Annual Report.

Naresh Kumar, S. 2008b. Carbon sequestration in coconut plantations In: Climate Change and Indian Agriculture. Aggarwal, P.K. (Ed.), ICAR, New Delhi.

Omae, H. and Takeda, Y. 2003. Modeling, winter dormancy of tea buds and simulation in Southern Japan. JARQ, *Japan Agricultural Research Quarterly* **37**:189-194.

Panda, R.K., Stephens, W. and Matthews, R. 2003. Modelling the influence of irrigation on the potential yield of tea (Camellia sinensis) in North-East India. *Experimental Agriculture.* **39:**260-67.

Peiris, T.S.G. 1991. Relationship between coconut yield and rainfall at Ratmalagara. *Annual Report, Coconut Research Institute Sri Lanka.*

Picini, A.G., Camargo MBP de, Ortolani, A.A., Fazuoli, L.C., Gallo P.B. and de Camargo MBP. 1999. Test and analysis of agrometeorological models for predicting coffee yield. *Bragantia.* **58:**157-70.

Thomas Grist, P. and Menz, K. 1995. Modelling rubber growth as a function of climate and soils. Imperata-Project-Paper -Improving-Smallholder-Farming-Systems-in-Imperata-Areas-of-Southeast-Asia. No. 6, ii 20 pp.

Van Ranst, E., Tang, H., Groenemans, R., Sinthurahat, S. and Ranst, E. Van. 1996. Application of fuzzy logic to land suitability for rubber production in peninsular Thailand. *Geoderma.* **70:**1-19.

Vijaya Kumar, K., Jacob Mathew, Amarnath, C.H., Nambiar, P.T.N., Jose, C.T. and Balakrishnan, T.K. 1988. Influence of weather on coconut yield. *In: Agrometeorology of Plantation Crops (Proceedings of the National Seminar),* pp.124-32. Prasada, Rao, G.S.L.H.V. and Nair, R.R. (Eds), Kerala Agricultural University Press, Mannuthy.

Yang, J. Zhao, R.F. and Lu, C.Y. 1993. Fuzzy comprehensive judgement and area division of tea (*Camellia sinensis*) climatic elements in Yunnan Province. *Journal of Yunnan Agricultural University* **8:**25-30.

Zuidema, P.A., Gerritsma, W., Mommer, L. and Leffelaar, P.A. 2003. A physiological production model for cocoa: Model description and technical programme manual of CASE2 version 2.2. Wageningen University, The Netherlands 145 p.

3

An Assessment of Potential Impacts of Climate Change on Fruit Crops

R. H. Laxman, K. S. Shivashankara and N. K. Srinivasa Rao

DIVERSE climatic conditions available across the country provide ample opportunity to grow almost all types of horticultural crops and India is the second largest producer of fruits with a share of 10.9% in the global production after China (16.7%). India is the largest producer of banana, mango and papaya, and have attained highest productivity in grape, sapota and banana. The increasing population and also awareness about the health and nutritional benefits of fruits among the urban consumers has increased the demand by manifold. To meet this demand, the production and productivity need to be further enhanced by exploiting the suitability of agro-ecological zones for different fruit crops.

Variability in weather conditions, like increase in temperature and moisture stress conditions, during critical stages of crop cause heavy yield loss and adversely affect the fruit quality even under adequate management conditions. There is enough scientific evidence to conclude that climate change is happening. Inter Governmental Panel on Climate Change (IPCC) estimated that the current global mean surface temperature was about 0.42–0.54 °C above the 1961-90 annual average (IPCC, 2007). It also ranks eleventh of the last 12 years (1995–2006) among the 12 warmest years. Due to climate change the increase in temperature and variability in precipitation pattern could lead to the development of abiotic stresses like heat, drought and flooding. These stresses could occur in varying intensities coinciding with different growth and developmental phases of the crops, finally determining the productivity and quality. An assessment of the impacts of elevated temperature and CO_2 on different perennial fruit crops and appropriate adaptation measures to overcome the negative impacts of climate change and future research priorities are discussed hereunder.

IMPACT ANALYSIS ON FRUIT CROPS

Elevated CO_2 and Temperature

Productivity

Carbon dioxide plays a central role in physiology of plants and there is a strong consensus in scientific community that elevated atmospheric CO_2 increases the rate of photosynthesis in C3 plants. The transpiration efficiency of leaves of both C3 and C4 plants increases and concomitantly decreases the amount of water lost through transpiration per unit of photosynthate produced leading to higher water-use efficiency. In tree crops, the canopy as a whole is essentially viewed as a source tissue and fruits are major sink. To get higher productivity, maximum allocation of resources has to be towards reproductive processes. The studies conducted at IIHR, Bangalore, under Open Top Chambers in grapes Thomposon Seedless showed that photosynthesis was higher 13.37 μ mol m^{-2} s^{-1} at 550 ppm CO_2 as compared to 10.43 μ mol m^{-2} s^{-1} at ambient levels. Further, studies are required to assess the impact of increase in 1 °C temperature and 550 ppm CO_2 and their interactions on different cultivars of grape in order to assess the impact of climate change.

Perennial fruit crops also have their own temperature optima and threshold levels and respond differentially for variation in temperature during growth and development. Some perennial fruit crops like pomegranate, guava and citrus have been reported to withstand the variations in temperature conditions. With this basic understanding further studies need be taken up to effectively use such crops in areas, which are going to experience higher temperature conditions. As the studies on impact of increase in temperature and CO_2 on perennial fruit crops is scarce, the detailed studies are required to quantify the impact and assess the vulnerability of different perennial fruit crops to climate change conditions.

Global warming will have negative impact on tropical regions where prevailing temperatures are already high, however, in areas where cold temperature is the primary limiting factor the crop production will benefit from warming. The climate projection studies indicate a general increase in temperatures in the order of 3–6 °C over the base-period average, depending on the scenario, with more warming in northern parts than the southern parts of India (Lal *et al.*, 1995; Lonergan, 1998; ERI, 2001). The projected increases in temperature in the short-term are in the range of 1.0–1.5 °C. Hence, to prepare ourselves to face the challenges of climate change in the short term, we need to first assess the likely impact of climate change with an increase of 1 °C in temperature and 550 ppm increase in CO_2.

The response of perennial fruit crops to increase in temperature is different as compared to annual crops. The perennial crops have deep root system and undergo vegetative and reproductive phases during different seasons. An average increase of 1 °C could affect the phenology of crop by influencing the degree-days and different crops respond differentially to the increases in temperature.

Mango: In India, mango is an important fruit crop and its phenology is highly influenced by variations in temperature. Like most tropical and subtropical tree fruits, mango has a vegetative bias, which becomes stronger with increasing temperature. The temperature influences the number and size of leaves initiated in vegetative buds of mango. Temperature, particularly in subtropics, provides a strong environmental stimulus, which has been shown to have an important role in flowering of mango. The optimum temperature range for mango growth has been reported to be 24–30 °C and base temperature for panicle growth and shoot growth are 12.5 and 15 °C, respectively. Mango can tolerate air temperatures up to 48 °C, and cold temperatures limit crop production. In the absence of studies on the influence of increase in temperature of 1 °C under controlled conditions on growth and development of mango, the studies conducted under field conditions on influence of temperature on flowering pattern in mango are discussed.

The ratio between male and hermaphrodite flowers present on panicles is temperature dependent. Studies have shown that the panicles that emerged late on Dashehari mango trees had up to seven times the percentage of perfect flowers compared with early emerging panicles on the same trees. Mean maximum and minimum temperatures at the time of emergence were 20.7/3.7 and 27.3/13.1 °C for early and late emerging panicles, respectively. During subsequent growth and development leading to anthesis, temperatures were 24.6/2.9 and 26.5/10.3 °C for early and late emerging panicles, respectively (Singh *et al*., 1966). The percent hermaphrodite flowers in some south Indian cultivars was much lesser when they were grown in northern India, which was attributed to lower, maximum and minimum temperatures during the period of panicle development at Delhi compared with southern India (Singh *et al*., 1965).

The evaluation of south Indian cultivars indicated that the percentage of hermaphrodite flowers was greatest in late emerging panicles, which were coincident with higher temperatures (Ramswamy and Vijayakumar, 1992). Singh *et al*. (1998) reported that panicles, which develop during December-January, when environmental temperature is comparatively low, usually have higher proportion of male flowers and tend to be malformed. They strongly recommended thinning out all panicles, which appear in mango trees early in the season under north Indian conditions.

Sharma and Singh (2006) in their study on the effect of pruning intensity on microclimate modification, sex ratio, malformation incidence and number of fruited panicles in 'Amrapali' mango trees under high-density planting, showed that malformation incidence was highest (44.2%) in panicles that appeared in un-pruned trees and lowest (36.2%) in panicles that appeared in severely pruned trees. Similarly, panicles that appeared at tree top had lower malformation incidence (37.2%) than those which appeared at lower canopy heights. Further, highest fruited panicles appeared in moderately pruned trees (80.3) and lowest in un-pruned trees (24.6). Maximum fruited panicles appeared at canopy top (87.4), but their number was drastically reduced at lower canopy heights.

Pruning intensities modify microclimate, the temperature observed was lowest (31.3 and 25.3 °C) in un-pruned trees and highest (36.4 and 27.9 °C) in severely pruned trees in August and March, respectively. Mango malformation is widely spread in north Indian subtropical regions where temperature is quite low. In Amrapalli, which is susceptible to flower malformation, covering with polythene sheet before onset of flowering in first week of January showed that, flower bud differentiation was 12 days earlier, bud burst 14 days early, 50% flowering 19 days early, fruit setting 22 days early and total duration was 29 days early. The percent malformed panicles were 12.5 and 1.14% in uncovered and covered trees, respectively.

These are some of the studies with field observations made in India about the role of temperature on nature of mango flowering. The likely rise in average temperature of 1 °C during vegetative flushing and flowering of mango would cause changes in vegetative growth and flowering. As it is observed that higher temperatures tend to influence the bias towards vegetative growth in mango, the increase in temperature and also the increased CO_2 concentrations, could influence the number of vegetative flushes and subsequent maturity of shoots before the shoots receive congenial temperature for flower induction. The studies have also highlighted the influence of prevailing temperature on flower composition of mango panicles and subsequent fruit yield. The rise of 1 °C temperature under climate change situations could also alter the ratio of male to hermaphrodite flowers in mango more so under subtropical conditions of northern parts of India. If trees receive relatively higher temperature under climate change conditions compared to what they are receiving at present but sufficiently low to induce flowering, there could be higher number of hermaphrodite flowers compared to male flowers early in the season. This could also further enhance the fruit maturity leading to early harvesting. The incidence of mango malformation in North Indian condition, based on the observations made in the above-mentioned studies might be reduced with an average rise of 1 °C under climate change scenarios. However, detailed studies need to be taken up in different agro-ecological regions under field conditions to quantify the variations in phenological stages and impact of elevated CO_2 and temperature under controlled conditions.

Grape: Temperature plays an important role in perennial fruit crop grape. Though grape originated in temperate regions, in India it is grown in three distinct agro-ecological regions, namely subtropical, hot tropical and mild tropical climatic regions. The modifications in production systems allowed this crop to become adapted to environments markedly different to those at the centre of its origin. In viticulture, temperature summations, degree-days have been used as dependent variables in describing the timing of various grapevine phonological events and growth. The base temperature is 10°C and temperatures between 22 and 35 °C are most congenial. The variations in temperatures cause the alterations in developmental stages and ultimately the ripening time. Vine growth and productivity are dependent up on the assimilation of carbon via photosynthesis and subsequent translocation and allocation. The optimum temperature for photosynthesis is 22–32 °C depending upon the cultivars, beyond which it declines.

In a study on cv. Sharad Seedless conducted at Bangalore it was observed that, there were differences in phenology, and quality characteristics, due to the prevailing temperature conditions. Maximum number of days (14 days) for bud break were observed in 8 November pruned shoots where the weekly average maximum (26.6 °C and 27.7 °C) and minimum (17.3 °C and 14.8 °C) temperatures were lower during the following two weeks, respectively. Number of days taken for maturity was also higher (132 days) due to lower minimum temperature prevailing during December (12.9 °C), January (15.5 °C) and February (14.1 °C). The colour development was very good in early pruned vines, due to low temperatures prevailing during colour development as compared to late pruned ones. The lower temperatures extended the duration taken for different phenological events and higher temperature during colour development phase affected the berry colour (Laxman and Srinivasa Rao, 2005).

The analysis of average mean monthly maximum and minimum temperatures prevailing during growing season, after fore-pruning, in grape growing areas showed that maximum temperature prevailing in Bangalore is lowest among the regions followed by Nashik (Fig.3.1).

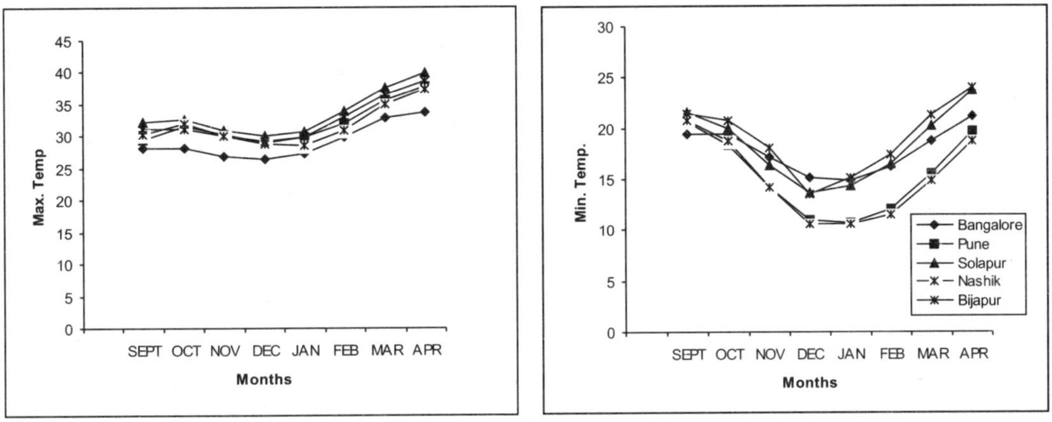

Fig. 3.1. Mean maximum and minimum temperatures prevailing during growing season in different grape-growing areas

Sholapur region has got highest maximum temperature. The lowest minimum temperatures are observed in Nashik and Pune regions. Bangalore and Sholapur regions showed relatively higher minimum temperatures.

The data on days taken for wine cultivar Cabernet Sauvignon for different phonological stages, viz., bud break, flowering, variation and harvest maturity collected for Nasik, Bangalore, Bijapur and Sangli areas showed that there are differences in days taken for variation and harvest maturity among grape-growing areas. Under Bangalore conditions, number of days taken for maturity is highest, followed by Nasik region and under Bijapur conditions lowest number of days are required for maturity.

Quality

Excessively high temperatures for extended periods of time generally result in a delay of fruit maturation and a reduction in fruit quality of grapes. The fruit quality is affected due to a decline in total titratable acidity and increased pH. High temperatures also reduce colour development in grape berries. The relationship between maximum, minimum and difference between maximum and minimum temperatures and quality parameters for cv. Cabernet Sauvignon were worked out (Anonymous, 2008). The relationship between mean maximum and minimum temperature during veraison to fruit maturity and berry anthocyanin content showed a negative relationship. The increase in both maximum and minimum temperatures caused reduction in berry anthocyanin content significantly at harvesting.

The difference between maximum and minimum temperature during fruit maturation was found to have very high influence on berry anthocyanin content. Higher the gap between maximum and minimum temperatures, higher the anthocyanin content. This relationship has greater bearing on the anthocyanin content under climate change conditions. Though both maximum and minimum temperatures would be increasing, the increases in minimum temperature would be higher, thereby reducing the difference between maximum and minimum temperature in turn affecting the quality. The relationships between mean maximum and minimum temperatures and berry total flavonoids content also showed similar trend.

The increase in both maximum and minimum temperatures caused the reduction in berry total flavonoids content. These relationships worked out between the prevailing mean maximum and minimum temperatures during fruit maturation phase (veraison to harvesting) have greater significance in assessing the impact of climate change on grape fruit quality. The increases in 1 °C temperature both in maximum and more so in minimum temperature would have considerable effect on quality parameters of grapes Cabernet Sauvingon.

Pests of Target Crop

The information on impact of elevated CO_2 and rise in temperature on pest of most fruit tree crops is lacking.

Aberrant Weather and Extreme Events

Observations on aberrant weather and extreme events on most of perennial fruit crops are not well documented and readily available.

Adaptation Measure for Climate Change and Global Warming

Using suitable cultivars: Adapting perennial fruit production systems to a changing climate has received little research attention worldwide. The vulnerability of any sector to climate change depends on both the expected regional climate change and sector's ability to adapt. The adaptation of fruit plantation to climate change takes time and requires long-term investments. The adoption of new varieties – a commonly cited option for climate change adaptation – occurs much more slowly than for annual crops. The long time horizon of perennial agriculture creates special challenges in a changing climate. Favourable areas may become unfavourable during the life of a single orchard or vineyard. The choice of a variety is complicated by the risk that the best variety for the current climate may be poorly suited for future climates. Thus, while adaptations such as planting new varieties and shifting to new areas may reduce impacts in long-term, short-term losses may largely be unavoidable. In situations, where there is a strong consumer preference for a select cultivar and also the suitable varieties are not available to adapt to the changing climate of a particular growing region, the option of using rootstocks for better performance of scion cultivars could be explored.

Alteration in crop choice: Another adaptation option could be the selection of appropriate fruit species or variety to suit to the changed climatic conditions. Certain physiological and morphological adaptations have endowed many fruit crops with the capacity to withstand adverse effects of water stress. Ber sheds leaves to avoid extreme water stress during summer. Fig has adopted to retain high bound water in the tissue, by having sunken stomata, thick cuticle and leaf wax coating. Ber, phalsa and tamarind also have sunken stomata, thick cuticle and wax coating of leaves. Indian gooseberry has adapted by reducing leaf area, thereby reducing the transpirational area. Pomegranate is fairly winter hardy and also drought tolerant. Aonla, being a hardy and drought tolerant subtropical tree, can be grown well under tropical conditions. These crops and many other fruit crops could be considered as candidate crops under climate change conditions.

Water management: The sound production strategies like improved water-use efficiency under hot and dry conditions should be developed. In addition to drip irrigation and mulching for production of fruit crops under water limiting conditions, novel irrigation methods, like partial root zone drying (PRD), could be adapted.

Mitigation Measure to Reduce Emission of CO_2/GHG and Carbon Sequestration Potential of the crop

Mitigation measures agriculture and forestry sectors are generating much interest as a potential source for additional income to otherwise weak rural areas and as a means of fueling adaptation to climate change. Tree fruit crops have substantial carbon sequestration potential and strengthen the system's ability to cope with adverse impacts of changing climatic conditions. Though there is scanty information available about carbon sequestration potential of fruit trees, they contribute significantly to climate change mitigation through enhanced carbon sequestration. In a study using PRO-COMAP model, Ravindranath *et al.* (2007) estimated the mitigation potential of farm forestry fruit orchard block planting with 75% of area proposed under *Mangifera indica*, *Tamarindus indica*, *Achras sapota*, *Artocarpus* and neem in Guavain the Kolar district of Karnataka. The carbon stock change under baseline and mitigation scenario (excluding harvested wood products) and carbon increment per ha for various project activities for 30-year period (2005–2035) worked out to be 47.42 tonnes C/ha. The overall mitigation potential for farm forestry with an area of 5,381 ha is 81,750 tonnes C. The fruit orchards farm forestry is found to be most profitable at 29.92 % IRR. Other mitigation measures are as follows.

- Resource conservation techniques
- Organic farming.

FUTURE STRATEGIES FOR RESEARCH

- Studies on physiology, phenology, growth, yield and quality of fruit crops to elevated temperature, CO_2 and excess and deficit water stress.
- Analysis and documentation of variations in phenology of fruit trees in relation to climate variabilities in different growing areas.
- Identification and development of crops and varieties tolerant to heat, water stress and climate aberrations for different agro-ecological regions and growing seasons.
- Identification of ideal rootstocks against biotic and abiotic stresses.
- Analysis of relationship between temperature and fruit colour development. Development of adaptation strategies by employing different crops, cultivars, plant architecture, crop management and protection techniques and post-harvest management under different climate change conditions.
- Development of crop simulation models to assess the impact of climate change on growth, yield and quality under different agro-ecological conditions.

CONCLUSION

In India, diverse climatic conditions, available across the country provide ample opportunity to grow almost all types of horticultural crops, thus making our country the second largest producer of fruits. The global climate change, due to increase in temperature and variability in precipitation pattern results in abiotic stresses like heat, drought and flooding. These stresses could occur in varying intensities coinciding with different growth and developmental phases of the crops affecting productivity and quality. An average increase of 1°C could affect the phenology of crop by influencing degree-days and different crops respond differentially to the increases in temperature. Understanding, the likely impacts of increase in temperature and CO_2 on fruit crops is the first step in developing sound adaptation strategies to address the adverse impacts of climate change. Hence, investigations on the effect of elevated temperature and carbon dioxide, drought and flooding on perennial fruit crops and the subsequent assessment of impacts of climate change on major fruit crops under different climate change scenarios is urgently required.

References

Anonymous. 2008. *Annual Report* ICAR Network Project on Impact, Adaptation and Vulnerability of Indian Agriculture to Climate Change for 2007-08.

Bindi, M., Fibbi, L. and Migliegetta, F. 2001. Free air CO_2 enrichment (FACE) of grapevine (*Vitis vinifera* L.) : II. Growth and quality of grape and wine in response to elevated CO_2 concentrations. *European Journal of Agronomy* **14:**145-55.

ERI, 2001. India's first national communication to UNFCCC. The Energy Resources Institute, New Delhi, India.

IPCC, 2007. Climate change 2007. The physical Science Basis. Contribution of working Group I to the Fourth Assessment Report of the Inter governmental Panel on climate change. Solomon, S., D. Qin, M. Manning, J., Chen, M. Manning, Z, Chen, M., Marquis, K. B, K.B. Averyt, M. Tignor and H.L. Miller (Eds). Cambridge University Press, Cambridge United Kingdom and New York, USA, 996 pp.

Lal, M., Cubasch, U., Voss, R. and Waszkewitz, J. 1995. Effect of transient increase in greenhouse gases and sulphate aerosols on monsoon climate. *Curr. Sci.* **69:**752-63.

Laxman, R.H. and Srinivasa Rao, N.K. 2005. Influence of temperature on phenology, yield and quality characteristics of grapes cv. Sharad Seedless. *National Seminar on Impact, Adaptation and Vulnerability of Horticultural Crops to Climate Change,* held on 9th December 2005 at IIHR, Bangalore.

Lonergan, S. 1998. Climate warming and India. In: Dinar A, Mendelsohn R, Evenson R *et al* (eds). Measuring the impact of climate change on Indian agriculture. World Bank Technical Paper No 402, Washington DC Mall R.K., Singh R., Gupta, A., Sinivasan, G., Rathore, L.S. 2006. Impact of climate change on Indian agriculture: A review. *Climate Change* **78:**445-78.

Ramaswamy, N. and Vijayakumar, M. 1992. Studies of the effects of flowering and fruiting behaviour of South Indian mango cultivars. *Fourth International Mango Symposium,* Miami Beach. FL., 47.

Ravindranath, N.H., Murthy, I.K., Sudha, P., Ramprasad, V., Nagendra, M.D.V., Sahana CA, Srivathsa KG and Khan H. 2007. Methodological issues in forestry mitigation projects: a case study of Kolar district. *Mitig Adapt Strat Glob Change.* **12:**1077-1109.

Sharma, R.R. and Room Singh. 2006. Pruning intensity modifies canopy microclimate, and influences sex ratio, malformation incidence and development of fruited panicles in 'Amrapali' mango (*Mangifera indica* L.). *Sci. Horti.* **109:**118-22.

Singh, R.N., Mujumder, P.K., and Sharma. D.K. 1965. Studies on the bearing behaviour of some South Indian varieties of mango (*Mangifera indica* L.) under North Indian conditions. *Trop. Agri.* **42:**171.

Singh, R.N., Mujumder, P.K. and Sharma, D.K. 1966. Sex-expression in mango (*Mangifera indica* L.) with reference to prevailing temperature. *Proc. Amer. Soc. Hort. Sci.* **89:**228.

Singh, V.K., Saini, J.P. and Misra, A.K. 1998. Mango malformation in relation to physiological parameters under elevated temperature. *Indian J. Plant Physiol.* **3:**231-33.

4

Impact of Climate Change on Mango and Tropical Fruit Crops

S. R. Bhriguvanshi

PLANT growth and development are governed by the environmental conditions. Thus, farming is intimately related to the prevailing weather conditions. Weather assumes significance in nearly in every phase of horticultural crops from preparation of the pits to harvesting and storage.

Weather is the single major limiting factor in fruit production. Successful farming needs appropriate decisions in respect of time of planting, sowing, transplanting, scheduling of irrigation, fertilizer application, etc., considering weather conditions. The air temperature, cold wave and frost are decisive factors in plant growth and development in subtropical fruits like mango, guava and papaya. A sound knowledge and understanding of the complex interaction between climatic factors and biological processes of plants are essential to scientific farming. Global climatic observations indicate rise in earth surface temperature by 0.6 °C since the 19th century, while CO_2 concentration in atmosphere is now about 31% higher than 200 years ago. If it continues to increase at the same rate, it could nearly double by 2035, leading to further global warming. The impact of global warming on tropical horticultural crops need to be assessed urgently to evolve adaptation strategies and identify possible opportunities.

IMPACT OF CLIMATE ON MANGO AND TROPICAL FRUITS

Elevated CO_2 and Temperature

Precise information on the effect of 1 °C rise in temperature over current climate at elevated CO_2 level of 550 ppm on productivity, quality and pest of mango and other horticultural crops, is lacking. However, the likely effect of global warming on mango and a few horticultural crops as compared to their optimal climatic requirements is described here.

Productivity

The optimal temperature for growth of mango has been reported to be 23.8–29.4 °C. However, it can tolerate temperatures as high as 48 °C during the period of fruit development and maturity. In fact, higher temperature during fruit development hastens maturity and improves fruit size and quality. Temperature influences the uptake and metabolism of mineral nutrients in plants. High calcium content in fruits has been related to longer post-harvest self-life. High N content is often associated with shorter post-harvest life of fruits.

Flowering in mango a month earlier at Pune than at Mumbai has been reported with more or less similar temperatures at both the places. Similarly in northern India, mango flowers at least 30 days earlier in eastern Uttar Pradesh, than in western Uttar Pradesh due to shorter winters and early onset of warming by a month in eastern Uttar Pradesh region. Prolonged spell of cloudy weather interferes with normal fruit setting. Entrapment of back radiated heat from the soil by the clouds is attributed for fruitlet and flower droppings in mango.

Temperature during inflorescence is crucial in mango. In controlled environment, low temperatures (15 °C day/10 °C night) during inflorescence, morphogenesis reduces the proportion of perfect flowers, the reduction being greatest in a tropically evolved polyembryonic group of cultivars compared with those from monoembryonic group. Cool temperatures during inflorescence development reduces the number of perfect flowers. Inflorescences which emerge during the middle and at end of the flowering season have been reported to produce between two and seven times more perfect flowers, respectively than the early breaking inflorescence. The increase in perfect flowers in the later emerging inflorescence correlates with higher average temperatures during the later part of flowering season.

Rainfall is crucial for mango: Well-distributed and not the total rainfall is of prime importance for growth and fruiting of mango. The monthly and annual rainfall patterns of important mango-growing countries indicate that there is very little rain, at the flowering time, which is desirable. Rains at flowering are harmful for mango causing total crop failure sometimes. Various workers have reported wash out of pollen-grains by rains during flowering from different parts of the world. The very delicate mango flower is easily injured in moist weather. During rainy days, pollinating insects remain dull, and effective pollination cannot occur. After continued moist weather, severe attacks of mango hoppers and certain fungi cause heavy shedding of flowers and fruits, and often result in crop failure. Pre-harvest low light intensity due to cloudy weather reduces content of ascorbic acid and sugar in fruits.

The mango can do well in areas having mean annual rainfall of as low as 25 cm to as high as 250 cm. However, it suffers severely in areas experiencing frequent rains or high humidity (above 80%) during flowering period. Such conditions are not conducive to a good fruit setting and increase the incidence of serious pests and diseases like powdery mildew and anthracnose.

Quality

Information on effect of elevated CO_2 and temperature on quality of mango and other horticultural crops is lacking.

Pest of Target Crops

Powdery mildew: The powdery mildew (*Oidium mangiferae* Berthet) is a sporadic, but serious disease of mango inflorescence causing 80–90% of crop loss. High relative humidity of 60–90% and temperatures ranging from 10–31 °C are conducive for the disease. The period identified for the out break of powdery mildew in Kakori and Malihabad mango belt of Uttar Pradesh is third/fourth week of March.

Anthracnose: Temperatures ranging from 24.88–34.06 °C and relative humidity of 81.00–89.75% are more conducive for development of Anthracnose (*Collectotrichum gloeosporioiodes*) during August and September.

B. dorsalis: Pupal period of *B. dorsalis* is longest (18 days) at 15 °C and shorter (6 days) at 35 °C. Warm, humid weather is favourable for *Bactrocera* fruit flies and pest population builds up during mango ripening. *Bactrocera* populations decrease during day period. Mango, citrus, papaya, litchi and guava are the most preferred by the fly.

Lepidopteran pest complex: Twelve species of *Lepidopteran* pests infest mango trees at different growth stages of mango (cv. Neelum) orchard in Tirupati, Andhra Pradesh, India.

A host of other pests damage the mango crop. These are shoot-borer (*Penicillaria jocosastrix*), leaf miner (*Acrocercops syngramma*), leaf-eating caterpillars (*Euthalia garuda garuda* [*E. aconthelia*], *Porthesia scintillans* [*Eupoctris scintillans*], *Lymantria marginata*, *Latoia lepida* [*Parasa lepida*], *Orgyia postica* and *Thalassodes quadraria*), flower-feeding caterpillar (*Euproctis fraterna*) and leaf folder (*Homona permutata*) which damage the new vegetative flush and flowers from July — March. However, leaf webber (*Orthaga exvinacea*) and fruit-borer (*Conogethes punctiferalis*) infest mainly the old foliage or matured leaves (September-October) and ripe fruits (April-June). Among the insects, bark-eating caterpillar (*Indarbela tetraonis*) damage mango crop for a longer period (July-January). The leaf gall insects, fruit flies, termite and thrips are active during second week of September, and last week of May, February and January, respectively.

Most of these pests and diseases are positively correlated with climatic factors of rains, high humidity and moderate temperatures.

Aberrant Weather and Extreme Events

Unusual impact of climate change has been witnessed in Orissa. Exceptional rise in daily maximum and minimum temperatures by about 1-2 °C above the normal temperatures in the coastal areas known for its moderate temperature, is now experiencing a mean maximum temperature of well above 40 °C, not witnessed earlier. This has changed the cropping/flowering patterns. Mango and mahua trees have been reported to flower unusually early. Normally mango trees in Orissa generally begin to flower in November, while mahua trees flowers from February. From 2000, the trend has been changing, with mango trees flowering in September and mahua in December. Early flowering normally leads to low yield and this is what Orissa has been experiencing from 2000. Climate change is attributed for early flowering in mango and mahua.

At Central Institute for Subtropical Horticulture, Lucknow, low temperature (4-11.5 °C) in January, 2003 accompanied by high humidity (>80%) and cloudy weather delayed panicle emergence. Temperature influences fruit setting too. Maximum fruit setting is observed on panicles which emerge first. Although some fruits set on the panicle emerged subsequently but they are dropped due to sudden increase in temperature (9–33 °C) and high wind velocity after Holi festival in March. The characteristics "Jhumka" (embryo abortion) with aborted embryos is noticed in and around Lucknow, Malihabad, Kakori and Mohan areas. The Directorate of

Horticulture, Uttar Pradesh, estimated the losses to be 30%. during 2005. The state experienced low fruit setting in mango due to sudden rise in temperature (35 °C) from 13 March onward in year 2005 as well.

Fruit crops including mango are sensitive to abrupt changes in weather. The high temperature accompanied by low relative humidity (49%) and long duration of sunshine hours (7.3-10.10 hours per day) resulted in excessive transpiration (6.0-10.0 mm per day) from the plants severely damaging the bloom owing to tissue dehydration. A temperature of 44 °C with 15% humidity in shade affected the young mango leaves and fruitlets in Palestine. Some of the fruitlets dropped, while others which survived and developed were without seed formation.

Mango is sensitive to frosting. Crops were affected due to cold wave caused on 9 and 10 January, 2006 when temperature went down as low as -0.5 °C in and around Lucknow. The extent of damage varied from 50 to 60% in young and 20–50% in old trees of mango. The damage was observed in fruits other than mango like guava, aonla, banana, papaya, bael, karonda, chironji, khirni, mahua, tamarind, wood apple and jamun.

At the Fruit Research Station, Saharanpur, Uttar Pradesh a temperature of -0.5 °C and 0 °C on two consecutive days for duration of about 1 hour 15 minutes killed one-year-old plants in spite of thatched protection of mango trees, whereas four-year-old plants were partially damaged with killing of their growing tips. Well-established mango trees of 12-year of age were also affected to a marked extent. Varietal differences in susceptibility to frost injury in mango were noticed.

Adaptation Measures for Climate Change and Global Warming

- Avoid new plantations in likely frost-prone areas in future.
- Give irrigation or spray water to reduce frost damage in the field.
- Use heater in orghard to raise temperature as much as 5 °C against frost.
- Planting shelter-belt or shade trees on orchard boundaries to protect the fruit crops from mild frost is required.
- Cover the soil with suitable materials, like paper, plastic or cellophane to reduce the cooling of the ground.
- Protect newly planted mango plants with suitable thatched material.
- Forewarning of cold waves and frost by regional and state meteorological departments and agencies to the farmers for timely protective measures is necessary.

Mitigation Measures to Reduce Emission of CO_2/GHG and Carbon Sequestration Potential of Crops

Fruit trees being perennial in nature have high potential for carbon sequestration. However, the same has not been properly evaluated and the information is lacking, and needs to be evaluated urgently. Other mitigation measures are as follows.

- Resource conservation techniques
- Organic farming.

FUTURE STRATEGIES FOR RESEARCH

- Breeding heat, drought and frost tolerant varieties having resistance to pests and diseases of mango, guava and papaya.
- Development of desired genetically modified fruit plants.
- Quantitative assessment of crop responses to enhanced levels of GHG, UV-B radiation and rainfall.
- Projections of impacts of global warming and climate change on regional and national scale based on modelling approaches.
- Development of effective pre-and post-harvest management technologies to minimise spoilage in fruit crops.

CONCLUSION

Mango grows well in temperatures ranging from 23.8–29.4 °C. The temperature during inflorescence is crucial. Cool temperatures and frosting during inflorescence development reduce the number of perfect flowers drastically. However, it can tolerate temperatures as high as 48 °C during the period of fruit development and maturity. Higher temperature during fruit development hastens maturity and improves fruit size and quality. Prolonged spell of cloudy weather interferes with normal fruit setting and is attributed for droppings of fruitlets and flowers. Therefore, well-distributed and not the total rainfall is of prime importance for growth and fruiting of mango. Rains at flowering are harmful for mango causing total crop failure sometimes. Wash out of pollen, flower injury and diminished activity of pollinators hinders fruiting. Additionally, moist weather favours severe attacks of mango hoppers and certain fungi causing heavy shedding of flowers and fruits, and may aggravate diseases like powdery mildew and anthracnose. Pre-harvest low light intensity due to cloudy weather reduces content of ascorbic acid and sugar in fruits. Under the impact of future climate change and global warming mango will be affected by extreme events of unseasonal erratic rains and frosting. Information on effect of elevated CO_2 and rise in temperature is lacking, which needs to be generated urgently to make projections on regional and national levels.

5

Impact of Climate Change on Oil Palm in India

K. Suresh and M. Kochu Babu

CLIMATE change refers to the increase in greenhouse gas emissions from burning of fossil fuels mainly from industrial activities and motor transportation. It is predicted that the area averaged annual mean warming would be about 3 °C in 2050's and about 5 °C in 2080's over the land regions of Asia. The atmospheric concentration of CO_2 fluctuated between 200 ppm during ice age and 270 ppm during warm intervals. The atmospheric CO_2 was roughly 270 ppm during the beginning of the Industrial revolution. The 377 ppm registered during 2004 may be nearly a level not seen for the last 55 million years. Since 1970, the earth's average temperature has risen by 0.8 °C and the rise in temperature each decade has been more than that of the preceding one. Meteorologists opine that the 22 warmest years recorded have come since 1980 and the six warmest years since 1880 have come in the last 8 years. Three of these six years, i.e., 2002, 2003 and 2005, were in which the major food-producing regions saw their crops wither in the face of record temperature. In India, there has been an increase of approximately 0.68 °C per century and an increasing trend in annual mean temperature with warming being more pronounced during post monsoon and winter.

CLIMATE CHANGE ANALYSIS IN OIL PALM-GROWING AREAS

The mean temperature in coastal Andhra Pradesh (East Coast) and Kerala (West Coast) from 1901û2006 reveals an increase of 0.8 °C and 1.2 °C, respectively. The rainfall pattern from 1901–2006 shows an increase of 3.47% in the coastal Andhra Pradesh (East Coast) and 2.58% in Kerala (West Coast). Mean temperature data in Andhra Pradesh indicates that there has been an increase of 1.6 °C and 1.59 °C in East and West Godavari districts, respectively during 1971–2002.

Effect of Elevated CO_2 and Temperature

Productivity

The studies on performance of oil palm under elevated CO_2 environments of climate change in India are lacking. However, oil palm being a C_3 plant, there are chances of increased photosynthetic rates and decreased photorespiration, which might manifest in higher leaf area, dry matter production and FFB yields. Observations on oil palm seedlings in CO_2 enriched indicated an increased photosynthetic rate (12 fold) due to higher intercellular CO_2 level (Ci), increased stomatal conductance and transpiration (3 fold), and water-use- efficiency (4 fold) as compared to the control.

The ideal climatic requirements for oil palm grown under rainfed conditions are as follows:

- Annual rainfall of 2000 mm or more evenly distributed, without a marked dry season, and preferably at least 100 mm in each month.
- A mean maximum temperature of about 29-33 °C and a mean minimum temperature of about 22–24 °C.
- Sunshine of 5–7 hours/day in all months and solar radiation of 15 MJ/m^2/day.
- Relative humidity above 45%.
- Low vapour pressure deficit.

It is difficult to separate the effect of minimum and maximum temperature on growth and yield of oil palm. The optimum mean temperature range is 24–28 °C, although palms at high elevation or at the geographical limit of about 15 °N may be growing at mean minimum temperatures <20 °C in part of the year. As temperature rises, photosynthetic activity in oil palm increases until the temperature reaches 20 °C. The rate of photosynthesis then plateaus until the temperature reaches 35 °C, where it begins to declines and at 40 °C, photosynthesis ceases entirely.

Oil palm, being an equatorial crop requires high temperature for its growth and yield. The effect of rise in temperature due to global warming, on performance of oil palm in India is lacking. Phytotron studies indicate that the best growth was at 32/22 (mean 27 °C). The next level with a mean temperature of 22 °C gave only slightly slower growth. At a mean temperature of 17 °C, it was only half of the best and very little growth occurred at a mean temperature of 12 °C (17/7). Higher temperatures can reduce or even halt photosynthesis, prevent pollination and anthesis, decrease the number of female inflorescences, abortion of bunches leading to bunch failure.

A brief history of cultivation of oil palm in India is available. Its performances in terms of FFB yields are available from different existing. These plantations were done during 1988-89 by 86 farmers in Pedavegi and Lingapalem mandals of West Godavari district. Plantations of M/s Navabharat Enterprises in over 200 ha were done during 1988–91. Department of Biotechnology also did plantations in collaboration with the Governments of Andhra Pradesh, Karnataka and Maharashtra in 1,000 ha in each state during 1990-91.The FFB yields from these oil palm plantations are available from late nineties only. About 80% of area under oil palm in India is in Andhra Pradesh. East and West Godavari district along with Krishna lead from the forefront in area, production and productivity of oil palm.

Crop simulation models for oil palm are not available for Indian conditions. Therefore, correlation between FFB yields and mean temperatures were worked out in major oil palm-growing areas of East and West Godavari districts of Andhra Pradesh and OPIL plantations of Kerala. The mean temperatures and FFB yields of oil palm plantations in East Godavari district of Andhra Pradesh from 1998–2007 (Table 5.1) indicates that there has been an increase of

0.3 °C increase in temperature during the above period. However, FFB yields did not vary among different years, indicating that mean air temperature had little effect on it. As oil palm plantations in Andhra Pradesh are mostly grown under irrigated conditions, management of nutrients (fertilizers) and water (irrigation) would be the key determining factors for FFB yields. Similar results were obtained for oil palm plantations in West Godavari district of Andhra Pradesh from 1999–2007 (Table 5.2). Data indicated an increase of 0.61 °C during the period, but the correlation between mean temperature and FFB yields were very poor. The increase in FFB yields over the years is mainly due to better management of water and nutrients. Similar results were obtained from rainfed oil palm plantations (OPIL) from Kerala showing very poor correlation between mean temperature and FEB yields.

Table 5.1. Mean temperatures and FFB yield (6 years after planting) at farmers' plantations in East Godavari district of Andhra Pradesh from 1998 to 2007

Year	*Mean temperature (°C)*			*FF Byield (tonnes/ha)*			*Average*
1998	28.6	17.5	19.5	12.5	15.0	12.8	15.4
1999	28.2	20.0	22.2	13.0	17.5	17.5	18.0
2000	28.1	18.8	17.0	13.5	21.3	18.8	17.9
2001	28.4	19.5	22.0	15.0	21.5	20.0	19.6
2002	28.4	22.0	23.8	17.5	22.3	22.5	21.6
2003	28.6	24.5	19.8	18.8	24.8	23.1	22.2
2004	28.8	17.0	14.5	21.3	25.8	23.6	20.4
2005	28.8	19.5	17.0	23.8	23.3	25.0	21.7
2006	28.8	19.8	24.5	25.5	25.3	25.3	24.1
2007	28.9	22.3	22.0	24.5	23.5	23.8	23.2

Table 5.2. Mean temperatures (mean T, °C) and oil palm FFB yields (6 years after planting) in West Godavari district of Andhra Pradesh from 1988 to 2007

Year	*Mean temperature*		*FFB yield (tonnes/ha)*		*Average*
1999	28.2	14.0	25.7	21.0	20.2
2000	28.1	14.9	21.5	26.2	20.9
2001	28.4	14.2	26.0	18.3	19.5
2002	28.5	14.0	21.8	17.7	17.8
2003	28.6	12.8	26.3	20.3	19.8
2004	28.8	13.8	28.7	33.8	25.4
2005	28.8	9.7	26.5	30.4	22.2

Quality

Higher temperature (>38 °C) during summer (1-2 weeks) might cause early ripening of bunches in the same whorl of palm (pseudo-ripening) leading to drastic reduction in oil extraction rates (<8%).

Pests of Target Crop

Warmer winter temperatures favor the increase of insect populations in oil palm. Higher temperatures may lead to more incidence of slug caterpillar, which causes defoliation of leaves and bunch end rot (25–50% loss). There is a negative correlation between the incidence of spear rot disease and average maximum temperature in Kerala under rainfed conditions. Overall temperature increases may also influence crop pathogen interactions by speeding up the pathogen growth rates, which increases reproductive generations per cycle, decreasing pathogen mortality rate due to warmer temperatures and by making the crop more vulnerable.

Aberrant Weather and Extreme Events

High temperatures dehydrate oil palm. When oil palm folds its leaflets to reduce exposure to the sun, photosynthesis is reduced. When the stomata on the abaxial side of the leaflets close to reduce the transpiration load, CO_2 uptake is also reduced and thereby photosynthesis.

- The effect of any kind of stress (temperature/water deficits) is revealed 1.5-2 years later on oil palm bunch yield.
- High temperature (>38 °C) leads to decreased oil extraction ratios by hastening the ripening of bunches (pseudo-ripening). However, such conditions may occur only during peak summer and may last for a week or so.
- Incidence of slug caterpillar on oil palm occurs during the summer months (April-May) in Andhra Pradesh, causing severe defoliation and reduction in yields
- Incidence of bunch end rot is noticed during summer, wherein a group of distal fruits abort and are delineated from the basal portion of bunch. Normally, 25–50% of the bunch is affected.
- Correlation studies of weather parameters with the manifestation of spear rot disease in oil palm during 1985–92 indicated a positive correlation of the disease with rainfall and number of rainy days, and a negative relationship with average maximum temperature.
- Higher incidences of spear rot and bud rot diseases were noticed during the rainy seasons.

Adaptation Measures for Climate Change and Global Warming

- Development of heat/drought tolerant hybrids.
- Development of disease and pest tolerant hybrids.
- Improved water and nutrient management techniques.
- Increasing the pollination efficiency of weevils.
- Improved access to timely weather updates/forecasts.
- Measures to reduce vulnerability to climatic risks like crop insurance and credit to farmers.

Mitigation Measures to Reduce Emission of CO_2/GHG and Carbon Sequestration Potential of Crops

Carbon Sequestration Potential of Oil Palm

- Perennial crops like oil palm serves as net accumulators of carbon, thereby offsetting carbon emissions, arising mainly from fossil fuel consumption. With respect to the net fixation or sequestration of carbon, there are possibilities of financial gains by developing countries like India through carbon trading, under the terms of Kyoto Protocol.

- Global carbon storage by oil palm is estimated at 74 Mt C/year for 12 m ha due to its high annual biomass productivity (50 tonnes DM/ha) compared to that of coconut (28 tonnes DM/ha/year).
- The amount of carbon sequestered by 11-year old oil palm hybrids under irrigated conditions ranged between 17.98 and 35.44 T C/ha with Papua New Guinea and Ivory Coast hybrids sequestering the highest and lowest carbon contents, respectively. Among the Palode hybrids, the carbon sequestered ranged from 20.26 (P12 × 313) to 26.05 T C.ha (P12 × 266). The carbon sequestered by the ASD Costa Rican hybrids was between 20.54 (ASD Deli × Ghana) and 35.44 T C/ha (ASD Deli × Ekona). The two Ivory Coast hybrids sequestered carbon between 17.98 and 28.73 T C/ha.

The following mitigation measures can be taken up in the form of Clean Development Mechanism (CDM) in case of oil palm:

- ***Composting:*** Composting of Empty Fruit Bunches (EFB) and Palm Oil Mill Effluent (POME) enables the mills to achieve zero waste management while at the same time producing organic compost could successfully supplement chemical fertilizer (partially) in the plantations.
- ***Biomass for energy production***: The oil palm biomass is used as fuels for boilers to generate power. This would displace fossil fuel based energy in mills. Palm oil can be used either as palm bio-diesel or as burning palm oil directly as bio-fuel to reduce fossil fuel to reduce carbon dioxide emissions.
- ***Forestry project:*** Oil palm plantations can also qualify as CDM forestry project. To qualify as a CDM project, projects must be large and are expected to save millions of tonnes of greenhouse gases per year. Landmass of about 1,000 ha would need to be assembled for a project. In India, where land holdings are small, a number of plantations would need to be assembled to make it large.
- Another study under irrigated conditions indicated that carbon contents in different forms of a mature palm ranged from 0.413–1.314 kg. The carbon contents were low in younger leaves and did not show any pattern among middle and lower whorls.
- With more than 80,000 ha of plantings in India, oil palm seems to be a prime candidate for storing carbon and is also eligible for Clean Development Mechanism (CDM).

FUTURE STRATEGIES FOR RESEARCH

As oil palm has been introduced into India in the late eighties and the establishment of NRC for Oil Palm only during 1995, research on oil palm on these aspects are very scanty and hence there is an urgent need to focus more on the impact of climate change on basic physiological aspects of the crop. The specific research priorities for this crop are as follows:

- Understanding the basic physiological and ecological response of oil palm to elevated CO_2 and temperature.
- Collection and compilation of databases relating to oil palm yield and weather data in all oil palm-growing regions of India.
- Study the impact of extreme weather events due to floods, heat, pests and diseases.
- Identification of critical climatic thresholds for oil palm-growing regions in India.

- Carbon sequestration studies in oil palm.
- Development of crop simulation models and decision support system for oil palm.

CONCLUSION

Oil palm was introduced in India recently two decades ago to meet the vegetable oil demand to the growing Indian population. The NRC for Oil Palm was established in 1995. Therefore, research on oil palm on the aspects of climate change and global warming is scanty. The mean temperature range is 24–28 °C for optimum growth and yield. Photosynthetic activity in oil palm increases linearly until 20 °C, thereafter it plateaus to decline after 35 °C and ceases completely at 40 °C. Warmer winter temperatures favour the increase of insect populations in oil palm. Being a perennial crop it has large potential for carbon sequestration for mitigation of climate change and utilization of wastelands. The amount of carbon sequestered by 11-year-old oil palm hybrids under irrigated conditions ranged between 17.98 and 35.44 T C/ha. High temperatures (>38 °C) leads to decreased oil extraction ratios by hastening the ripening of bunches (pseudo-ripening). Incidence of slug caterpillar on oil palm occurs during summer months (April–May) in Andhra Pradesh, causing severe defoliation and reduction in yields. High temperature, heavy rainfall and more number of rainy days are associated with disorders like bunch end rot, and spear rot and bud rot diseases.

6

Impact Analysis of Temperature Rise and Elevated Carbon Dioxide on Banana Production in India

I. Ravi and M. M. Mustaffa

CLIMATE change is a definite and proven fact. It is frequently assumed that global change will bring higher temperatures, altered precipitation, and higher levels of atmospheric CO_2 (IPCC, 2007). In India more than one billion people, about 68% are directly or indirectly involved in agriculture sector and this sector is particularly vulnerable to present-day climate variability (O'Brien *et al.*, 2004). Although, direct effects of climate change in terms of rise in temperature and elevated CO_2 may lead to productivity increases in some irrigated crops (Aggarwal and Mall, 2002), there is general consensus that agricultural production in most areas is likely to be adversely affected, particularly in areas to become increasingly water-stressed (Gadgil, 1995; Dinar, *et al.,* 1998; Kumar and Parikh, 2001; Lal *et al.*, 1998).

Globally, banana (*Musa* sp.) is fourth most important commodity after rice, wheat and corn. It is grown in more than 130 countries across the world in an area of 8.25 million ha producing 97.38 million tonnes of banana and plantain. India is the largest producer of banana in the world, contributing 24% to the global production of banana with a total production of 19.19 million tonnes from an area of 0.565 million ha. Among horticultural crops, contribution of banana to agricultural Gross Domestic Product (GDP) is highest (1.99%) (Singh, 2007). Banana a plant of tropics and subtropics, requires hot and humid climate. The most suitable climate for banana growth is one with warm moist weather throughout the year without strong winds. In India, banana is grown in the regions from the humid tropics to humid subtropics and semi-arid tropics and from the sea-level up to an elevation of 2000 MSL. Favourable factors for banana production are rainfall >100 mm per month (Simmonds, 1966) and a temperature range of 10–40 °C with an optimum temperature of 25–30 °C and a mean minimum temperature of 15.5 °C. Basically, bananas are water-loving plants; and large areas of bananas are grown along the river basins, *viz.* Gangetic plains of Bihar, around Godavari river basin in Andhra

Pradesh, areas around Tapti river in Maharashtra and Cauvery delta regions of Tamil Nadu and in other areas where water sources have been created. Fairly good soil drainage is essential, which favours growth of the banana crop. Broadly speaking, banana cultivation is widely practiced in areas with the 102 cm isohyet and the 15.5 °C (60 °F) winter isotherms which roughly coincides with 30° latitude N and S of the equator.

IMPACT ANALYSIS ON BANANA

Elevated CO_2 and Temperature

Productivity

Elevated CO_2: Schaffer *et al.* (1996) investigated effects of atmospheric CO_2 enrichment and root restriction on photosynthetic characteristics and growth of banana (*Musa* sp. AAA cv. Gros Michel) plants. In their investigation banana plants were grown aeroponically in root chambers in controlled environment glasshouse rooms at CO_2 concentrations of 350 or 1000 ìmol CO_2 mol^{-1}. At each CO_2 concentration, plants grown in large (200 litres) root chambers did not restrict root growth and in small (20 litres) root chambers restricted root growth. Plants grown at 350 ìmol CO_2 mol^{-1} generally had higher carboxylation efficiency than plants grown at 1000 ìmol CO_2 mol^{-1} although actual net CO_2 assimilation was higher at the higher ambient CO_2 concentration due to increased intercellular CO_2 concentrations resulting from CO_2 enrichment. The banana plants grown at 1000 ìmol CO_2 mol^{-1} accumulated more leaf area and dry weight than those grown at 350 ìmol CO_2 mol^{-1}. Plants grown in the large root chambers were more photosynthetically efficient than plants those grown in small root chambers. At 350 ìmol CO_2 mol^{-1}, leaf area and dry weights of plant organs were generally more for plants in large root chambers compared to those in the small root chambers. Atmospheric CO_2 enrichment may have compensated for effects of root restriction on plant growth since at 1000 ìmol CO_2 mol^{-1} there was generally no effect of root chamber size on plant dry weight. Therefore, higher CO_2 environment is having definite advantage in banana plant growth and production as evidenced by the above investigation. However, restriction on banana plant growth imposed by temperature rise of up to 1 °C over the current levels may be compensated through CO_2 fertilization effect.

Temperature: The optimum temperature for banana leaf production and plant growth is 32 °C, but leaf emergence stops below 10 °C. During summer, each plant may produce 4 or 5 leaves a month but in winter only about half a leaf per month is produced (Turner, 2003). A banana plant produces 40–50 leaves before flowering (Cavendish group) and it varies with genotypes. The rate of leaf production indicates the healthy status of plant as well as favourable climate. In a healthy banana plant, under favourable tropical conditions, a leaf is produced in 7-10 days intervals in "Dwarf Cavendish" bananas. But in winter, it takes 20 days for a leaf production. The different stages of leaf unfurling are completed in 4 days in summer, while in winter it takes additional 14 days or more. The rate of leaf production increases significantly with increase in mean monthly temperature. The rate of appearance of new leaves and rate of fruit growth are largely governed by temperature. The temperature has overriding effect on leaf production in different seasons (Turner, 1971). Leaf production increases by one leaf per month for every 3.3-3.7 °C rise in minimum or mean temperature from 10–20 °C or 13.5–25 °C, respectively (Turner, 1971; Robinson, 1981).

Leaves of banana plants are upright under cool conditions, but under warmer conditions, they are more horizontal. The rate of plant development (for any given variety) is slower in the subtropics than in the tropics, and maximum yields occur, in the present climate, at latitudes between 15° and 20°. The optimum temperature for bunch development is about 10 °C lesthan the optimum temperature for the rate of leaf appearance. For 'Williams' (Cavendish group) variety, optimum temperature for bunch growth is about 21 or 22 °C. Temperature has a big influence on the rate of fruit growth, thus use of bunch covers, which are thought to warm the fruit, thereby increased the growth rate (Turner *et al.*, 2007). Choking can also be caused by high temperatures (above 30 °C) and drought (Stover and Simmonds, 1972). Choking is a condition, when the distance between the petioles (leaf stalks) of alternate leaves is reduced and the plant takes on a "choked" or "rosette" appearance.

The average temperature may increase over current temperature and the shift in climatic zones will be more pronounced in higher latitudes. In mid latitudes, the shift is expected to 200-300 km for every increase in 1 °C (IUCC, 1992). In banana-growing states, Tamil Nadu, Kerala, Andhra Pradesh and South Karnataka coming under lower latitudinal regions of the country, the temperature rise may be lesser than mid and higher latitudinal regions.

The temperature rise by 1 °C across India may not affect banana production as the temperatures would be well within optimum temperature for banana growth and production (Table 6.1). These traditional banana-growing regions still favour the growth, development and yield of banana. The global circulation model also projected that the temperature rise around 1-2 °C in the subtropical regions of the country and may have positive influence on banana growth and development in India. This will especially favours the banana production in Maharashtra, Gujarat, Uttar Pradesh, Bihar, Jharkhand, West Bengal, Orissa and North-Eastern regions. The banana cultivation area may extend in these regions and further augmenting the production. In the banana cultivation areas of central and northern parts of the country will become favourable for banana growth and development and yield may increase as compared to present level of production. There is a possibility that more areas can be brought under banana cultivation in subtropical conditions under climate change.

Table 6.1. Average maximum and minimum temperature in different banana-growing areas of India during 2005-06

State	*Areas*	*Maximum (°C)*	*Minimum (°C)*
Tamil Nadu	Coimbatore	31.42	21.30
Tamil Nadu	Trichy	33.99	24.60
Karnataka	Arabhhavi	28.06	20.43
Maharashtra	Akola	33.32	19.74
Maharashtra	Jalgaon	34.64	19.88
Gujarat	Gandevi	33.41	19.03
Punjab	Ludhiana	29.52	17.52
Bihar	Pusa	27.81	20.03
West Bengal	Mohanpur	32.77	20.61
Assam	Jorhat	28.48	19.38

Source: AICRP – TF, 2004-05

Frost injury: Low temperatures in banana plantation can cause chilling symptoms on leaves, which are not seen immediately but it may take 2–4 days to appear. The lamina turns yellow and on the emerging leaf, especially in plant crops, the most recently emerged part of the midrib may show brown areas which are water-soaked underneath. On older leaves, symptoms can be similar to potash deficiency. Once temperatures drop below freezing, damage occurs quickly and a few hours later, the leaves and bunches become water-soaked, then blacken and die. Frost rarely kills the whole plant. It is usually only the exposed tissues which are damaged. Within a few days, the youngest leaf begins to emerge again and growth resumes. When plants are completely defoliated by frost, the formation of the bunch can be delayed and, if a bunch is already present in pseudostem, the yield will be reduced. These effects are not seen until some months after the frost has occurred (Turner, 2003). The bunch emerging during winter months (December-January) in sup-tropical regions, produce uneven and often deformed fruits. The fruit is tapered, being thin at the stalk end, and the flower end may have a conical, green protuberance. Often the fruits are short and thick when mature (Turner, 2003). The quality of fruits is good and has a reputation for being the best tasting of all bananas. When the distance between the petioles (leaf stalks) of alternate leaves is reduced, the plant takes on a "choked" or "rosette" appearance. This is commonly seen during late winter to early spring. The bunches may fail to emerge properly. This is common in Dwarf Cavendish but not so common in Williams variety. Choked bunches are prone to sunburn and produce fruits difficult to pack.

Quality

Information on effect of elevated CO_2 and rise in temperature on quality of banana is lacking.

Pests of Target Crop

Altered temperature, rainfall pattern and humidity have an impact on development and distribution of banana insect pests and diseases. Increase in temperature and CO_2 will lead to an increase in population of pests and severity of diseases in presence of host plants. It increases the rate of reproductive cycle of insects and pests. The increase in insect population leads to demand for more use of pesticides, harming the ecosystem. Incidence of pests and diseases would be most severe towards equatorial regions due to impact of climate change. Pests and diseases of plains ecosystem may gradually shift to hills and mountains as the temperature rises. The pests like banana weevils, aphids (vector of virus disease) may have more population in the changed environment. The leaf spot disease may be spread towards northern belt of India as higher temperature favours the disease development. At Jalgaon, the leaf spot diseases were not observed earlier, but of late leaf spot and Cucumber Mosaic Virus disease started appearing. These new developments may be due to climate change.

Trend Analysis of Banana Production in India

The banana production in Maharashtra, Andhra Pradesh, Karnataka, Assam, Bihar, Tamil Nadu, Kerala, and Gujarat increased substantially since 1982 (AICRP-TF, 1982, and 2004). The average maximum and minimum temperature fluctuations in these states showed an increasing trend ranging from 0.3–0.4 °C. However, banana production has increased tremendously. This may be both due to implementation of new technologies and synergistic effect of climate change. The studies conducted in AICRP –TTF programmes showed that at Coimbatore, Tamil Nadu the same Cavendish group banana yielded 12–18 and 28–34 kg/bunch in 1982 and 2004, respectively (Table 6.2). Increase in yield during the period was noticed for Palayankodan at Kannur, Kerala and Basrai Jalgaon, Maharashtra, also. It is concluded that till now the climate change over past 20 years did not have any negative impact on banana production.

Table 6.2. Performance of banana varieties in different states during 1982 and 2004

Variety	*States*	*Yield (kg/bunch) 1982*	*Yield (kg/bunch) 2004*
Cavendish group	Tamil Nadu	11 –18	28–34
Palyankodan	Kerala	7.14 – 8.12	11–12
Basrai	Jalgaon (Yawal)	11.13 – 14.4	22.37

Source: AICRP – TF – 1982-2004

Aberrant Weather and Extreme Events

At Jalgaon leaf spot diseases were not observed earlier, but of late leaf spot and Cucumber Mosaic Virus disease started appearing. These new developments may be due to climate change.

Adaptation Measures for Climate Change and Global Warming

Modification in agrmonomic cultural practices and fertilization to avoid drought and flooding. Using drought and flooding tolerant cultivars.

Measures to Reduce Emission of CO_2/GHG and Carbon Sequestration Potential of Crops

- Banana being a popular commercial crop of 12–15 months duration with huge biomass production could qualify for carbon sequestration and carbon trading scheme provided if we adopt perennial type of cultivation in banana. At a given point of time, certain amount of banana plant crop stand will remain in the field throughout the year even if harvested after maturity. Therefore, banana crop has the potential for carbon sequestration under perennial system of cultivation with zero tillage as adopted in hill regions of India. However, the potential has not been assessed properly.
- Resource conservation techniques.
- Organic farming.

FUTURE STRATEGIES FOR RESEARCH

- Simulation modeling for temperature rise and elevated CO_2 concentrations on banana growth and development.
- Evaluation of existing commercial varieties for its suitability to high and low temperature regions and identification of critical stages of temperature and development of improved management practices in such regions.
- Evaluation and development of banana varieties for soil moisture deficit stress and inland salinity.

- Waterlogging will become an important hypoxic stress in a heavily precipitated area and flood-prone area. Therefore, existing varieties are to be evaluated for water logging tolerance at different stages as well as improve the management practices in that environment.
- Relationship between temperature rise, and pests and diseases incidence has to be studied in different regions and development of forecasting model.

CONCLUSION

Banana production may not be affected severely in India, even if the temperature rise by 1 °C from the present level combined with elevated CO_2 concentrations. Since temperature is not a limiting factor for banana production, there is a possibility of increase in area under banana cultivation and production. Overall climate change will have positive impact on banana production. However, farmers will have to adapt to climate change through changes in farming practices, cropping patterns, and use of new technologies will help to ease the impact. The climate change is a slow process and the plants will also slowly acclimatize to new environment as it has been done over the centuries of evolution. Higher temperature, elevated CO_2 concentration, water deficit and higher incidence of pests and diseases are predicted under the climate change future scenarios. Therefore, there is a need to focus research in these areas to turn the negative impact of climate challenge into positive impact for banana production.

References

AICRP Tropical Fruits, *Annual Reports* (1982-2004).

Aggarwal, P.K. and Mall, R.K. 2002. Climate change and rice yields in diverse agro-environments of India. II. Effect of uncertainties in scenarios and crop models on impact assessment. *Climatic Change* **52:**331-43.

Dinar, A., Mendelsohn, R. and Songhi, A. 1998. Measuring the impact of climate change on Indian Agriculture. World Bank Technical Paper 402, Washington.

Gadgil, S. 1995. Climate change and agriculture: an Indian perspective. *Current Science* **9:**649-59.

IPCC, 2007, (Intergovernmental Panel on Climate Change). Climate Change 2007: Synthesis report. An Assessment of the Intergovernmental Panel on Climate Change. This summary, approved in detail at IPCC Plenary XXVII (Valencia, Spain, 12-17 November 2007), represents the formally agreed statement of the IPCC concerning key findings and uncertainties contained in the Working Group contributions to the Fourth Assessment Report. pp 27.

IUCC. 1992. Information unit on climate change. Fact Sheet no.101. Impacts of climate change. UNEP, Polaisdes Nations. CH 1211, Geneva.

Kumar, K.S.K. and Parikh, J. 2001. Indian agriculture and climate sensitivity. *Global Environmental Change* **11:**147-54.

Lal M, Singh KK, Rathore LS, Srinivasan G and Saseendran SA. 1998. Vulnerability of rice and wheat yields in north-west India to future changes in climate. *Agricultural and Forest Meteorology* **89:**101-14.

O'Brien, K.L., Leichenko, R., Kelkar, U., Venemad, H., Aandahla, G., Tompkinsa, H., Javedc, A., Bhadwalc, S.,Bargd, S., Nygaarda, L. and Westa, J. 2004. Mapping vulnerability to multiple stresses: climate change and globalization in India. *Global Environmental Change* **14:**303-13.

Robinson, J.C. 1981. Studies on the phenology and production potential of Williams banana in sub tropical climate. *Subtropics* **2:**12-16.

Schaffer, C., Searle, Whiley, A.W. and Nissen, R.J. 1996. Effects of atmospheric CO_2 enrichment and root restriction on leaf gas exchange and growth of banana (Musa). *Physiologia Plantarum* **97(4):**685-93.

Singh, H.P. 2007. Research and development of banana and plantain in India *In*: Souvenir. National Confrence on Banana Production and Utilization of Banana for Economic Livelihood and Nutritional Security. pp 1-19. October 25-28, 2007- Tiruchirapalli, Tamil Nadu.

Stover, R.H. and Simmonds, N.W. 1972. *Bananas*. Longman, London.

Turner, D.W. 1971. Effects of climate on rate of banana leaf growth *Trop. Agric.* (Trinidad) **48:** 283-87.

Turner, D.W. 2003. Bananas-response to temperature. NSW Agriculture, Australia Ag fact H6.2.6. pp 1-3.

Turner, D.W., Fortescue, J.A. and Thomas, D.S. 2007. Environmental physiology of the bananas (*Musa* spp.). *Braz. J. Plant Physiol.* **19:**463-84.

7

Effect of Climate Change on Yield and Insect Pests Incidence on Cashew

N. Yadukumar, T. N. Raviprasad and M. G. Bhat

CASHEW is extensively grown in 35 districts distributed in west coast, east coast and plateau regions in different states of India. It is mostly grown as a rainfed crop, therefore, its performance depends mainly on climate. Prolonged and unseasonal rainfall accompanied with high velocity wind during flushing and flowering period affects crop adversely as large proportion of lateral branches remain as non-flowering laterals. Drought conditions drastically reduce cashew nut production. Most of the cashew varieties are sensitive to salty seawater. Rise in sea water level under climate change conditions may adversely affect the cashew plantation. This chapter summarizes the likely impact of climate change on cashew production and productivity in India. Future adaptation and research strategies to overcome these effects on cashew are also given here.

IMPACT ANALYSIS ON CASHEW

Elevated CO_2 and Temperature

The information on effect of elevated CO_2 on cashew production is lacking. However, productivity is highly dependent on prevailing temperature, relative humidity and rainfall.

Productivity

Low productivity and production of cashew in India recently, despite increase in area under new cashew plantations and improved management practices may be attributed to climate change. Comparative studies of variation in production and productivity of rainfed cashew crop over the years clearly showed profound effect of climatic factors directly or indirectly. The total cashew production in Kannur and Kasaragod districts was poor during 1996-97 (32,000 tonnes) and 1997-98 (54,000 tonnes) compared to 1995-96 (70, 000 tonnes). It may be attributed to unfavourable weather conditions during 1996-97 and 1997-98. In 1996-97, moist weather due to moderate dew fall with mean minimum temperature about 20 °C during reproductive phase resulted in only moderate flowering and enhanced incidence of Tea Mosquito Bug (TMB) resulting in poor yield. On the other hand in 1997-98, very poor dew with slightly higher mean minimum temperatures of 21-22 °C during reproductive phase reduced flowering with consequent yield losses. In 1998-99 cashew production in Kannur and Kasaragod districts was lowest recorded till date though profuse flowering was obtained due to favourable weather during the reproductive phase. However, high dew fall in February 1999 led to unprecedented breakout of pests followed by heavy continuous rains in May 1999 during harvesting period drastically reduced the nut yield and quality in 1998-99 season. Evidence suggests that with scientific method of pest control measures are critical during and after peak flowering to improve yield.

The rainfed cashew crop is highly sensitive to changes in climate and weather, particularly during reproductive phase. Cashew genotypes markedly vary in their heat unit (day °C) requirement. Early variety like Anakkayam-1 requires only 1953 (day °C) for reproductive phase, while 2483 (day °C) are needed in case of late season types like Madakkathara-2. Heat units, cashew yield and kernel weight are positively correlated particularly for mid-and late maturing varieties. Delayed flowering and fruiting in late varieties always faces moisture stress, which can be managed by supplemental irrigation.

***Effect of temperature and relative humidity*:** Cashew requires relatively dry atmosphere and mild winter (15–20 °C minimum temperature) coupled with night with moderate dew for profuse flowering. High temperature (>34.4 °C) and low relative humidity (RH) of <20% during afternoon causes drying of flowers resulting in reduced yield. The mean maximum temperatures of 34.4, 36.1 and 36.2 °C were recorded during February 1992-93, 2003-04 and 2004-05, respectively. The corresponding RH in the afternoon was 33.0, 18.5 and 19.7%. Poor yields were obtained during 2003-04 and 2004-05 as compared to 1992-93 mainly due to drying of flowers.

***Effect of rainfall and dew on bud break and flowering in cashew*:** Continuous rains without critical dry spells and late winter rains delay the bud break in cashew. Normal favourable climate envisages early cessation of rains and intermittent dry spells and bright sunshine hours between August and September to trigger early bud break in cashew in west coast. Usually a dry spell of 7 days is necessary about 30 days prior to bud break. Late and extended winter rains reducing number of bright sunshine hours invariably delays bud break, better availability of soil moisture during flowering period (December and January). Incessant rains until November in 1998 delayed bud break. The delay in bud break was pronounced in early varieties. In late varieties, bud break was normal, because the required dry spell of 7 days was met 22–26 days before bud breaking.

Peak flowering was early (8 December) during normal years (1995-96) and late (29 December) in aberrant climate situation (1997-98) receiving late winter rains during November and December 1997. There was significant delay in peak flowering in Madakkathara -2 and Kanaka (13-1-98) during 1997-98, however, early varieties flowered on normal dates. The peak flowering outside the station was during January/February (delayed by one month), which may be attributed to poor crop management and planting material of seedlings of unknown origin.

Quality

The information on the effect of elevated CO_2 and temperature on quality of cashew is lacking and needs to be studied in detail to evolve adaptations to future climate change and global warming.

Pests of Target Crops

Prolonged and unseasonal rains as also heavy dew aggravate incidence of pests and diseases to the detriment of cashew yield and quality. The cashew flowered profusely during both 1995-96 and 1998-99 seasons but yields were severely affected in 1998-99 in Kannur and Kasaragod districts of Kerala due to unprecedented incidence of pests aggravated by heavy dew in February 1998 compared to less pestiferous 1995-96. Rains received during flowering and harvesting (2008) period increased the incidence of pests and diseases resulting in very poor yield and quality. Cashew suffers mainly from tea mosquito bug (*Helopeltis antonii* Signoret) and inflorescence blight disease. Incidence and severity of both are dependent on climate and weather factors.

Tea Mosquito Bug (TMB): Among many foliage infesting pests of cashew, tea mosquito bug (*Helopeltis antonii* Signoret) is most important in major cashew growing tracts of India. Its adults and nymphs feed on plant sap from the tender shoots, panicles and immature cashewnuts. The pest causes yield losses ranging from 30–90% depending on the extent of incidence. The pest population starts building up from September to October and gradually reaches a peak during February. The pest population is also synchronized with the phenological stages of the crop, which is also dependant on prevalent weather parameters. Various weather factors influence the pest build-up of which temperature and relative humidity have been reported to play a significant role.

A multiple regression prediction model was developed based on weather data and pest population from 1988 to 1998 with 88% accuracy of prediction of incidence.

The regression model $Y= 0.648\, x_1 + 0.0022\, x_2 + 0.048\, x_3 - 0.073\, x_4 - 0.642\, x_5 - 0.597\, x_6 - 0.272\, x_7 + 30.473$ was developed based on the pooled data of the TMB population.

wherein, Y= predicted pest population; x_1 = sunshine hours (hours/day); x_2 = rainfall (mm); x_3 = morning relative humidity (%); x_4 = evening relative humidity (%); x_5 = maximum temperature (°C); x_6 = minimum temperature (°C); x_7 = number of rainy days.

The model is helpful in predicting subsequent TMB population one month in advance based on the prevalent weather parameters. An increase in mean temperature to the tune of 1 °C can induce a reduction in the pest population up to 46% as indicated by the r value of 0.642 for x_5 (maximum temperature °C). However, favourable limits of temperature and RH need to be determined for this pest.

The minimum temperature plays a major role in incidence of pest population. The favourable minimum temperature for TMB incidence ranged between 13-18 °C. Low temperature (<12 °C) is antagonistic for pest build-up. Occurrence of cloudy weather favoured the rapid build-up of the pest as indicated by the observations recorded at Pilicode during 1990-91, 1993-94 and 1997-98.

***Inflorescence blight*:** The drying up of young shoots, inflorescence and immature nuts is caused by fungal pathogens, viz., *Colletotrichum gleosporoides* and *Gleosporium mangiferae*. The incidence of this disease is being reported from different new location wherein it was not prevalent earlier. High relative humidity in forenoon during December–February both 1997 and in 1998 and the minimum temperature of 18–20 °C were found to be conducive for sporulation by fungi. Further investigations are necessary for identifying the effect of rise in temperature on the occurrence of this disease. A significant increase in dewfall was one of the most important factors which favoured the growth, sporulation and spread of fungi. A lead-time of approximately one month was identified between conducive weather factors and incidence of disease (Prasada Rao, 2002). Cloudiness leading to low bright sunshine hours (< 2 hours/day) followed by dew/fall triggered growth, sporulation and spread of fungal pathogens causing inflorescence blight during 1998-99 in Kannur and Kasaragod district.

Aberrant Weather and Extreme Events

- ***Impact of Tsunami:*** The Tsunami in coastal Tamil Nadu on 26 December 2004 devastated coastal parts of Kanyakumari and Cuddalore districts. Tsunami severely affected on cashew crop. Most of the standing crop was inundated with salty seawater ingressing in the mainland. Surprisingly four cashew trees remained healthy and unaffected showing salt tolerance. The scions of trees were brought and grafted. These grafts were planted in germplasm plot for further studies.
- ***Impact of unseasonal heavy rains*:** In Dakshina Kannada and Udupi districts of Karnataka, and Kasaragod and Kannur districts of Kerala, nearly 200 mm unseasonal rainfall was received continuously from 15–25 March 2008, affecting cashew yield drastically. Peak harvesting in 2008 was over before 15 March 2008 and nuts were spoiled in field due to over absorption of rain water. Continuous rains did not permit drying nuts immediately resulting in poor quality. Large quantity of nuts, which could not be picked in time, germinated in the field itself. The nuts and apples at ripening stages were spoiled due to blackening of nuts as well as rotting of apple on trees itself. Flowers exposed to heavy rain accompanied with wind dropped off or attacked by fungus disease. Pollinating insect activity reduced to the minimum resulting in poor setting of nuts during that period. Total nut yield losses were 50% with heavy loss (65%) during peak harvesting period with very poor quality due to unseasonal rains received in March 2008. In contrast, during the previous two years (2005–07), no rains were received during the corresponding period with good yield. Normally cashew harvesting starts from January and lasts till May as in case of 2005-06 and 2006-07. Performance of the crop is good if continuous rain is not received until May allowing major harvesting to be over. A few summer rains are desirable for the crop but should not be in excess. Normally, quantity of harvested nuts is about 18% during February, reaching a peak of 55% in March and declining gradually to 26% of the total produce.

Adaptation Measures for Climate Change and Global Warming

- Under drought situations drying of flowers can be checked to some extent by adopting proper soil and water conservation measures. Formation of terrace and crescent bund and mulching the base area with cashew leaf litter and other *jungle* growth available in the garden are helpful.

- Supplemental irrigation of 200 litres of water/plant once in 15 days during January–March from water collected in ponds through rain harvesting. This helps in flowering and nut development by improving the microclimate with increased humidity, particularly in the afternoon. It also leads to increased nut and kernel weight by reducing flower and nut drying to some extent.
- Drip irrigation during fruit development stages wherever water is available may be helpful during the drought situations under climate change to rainfed cashew crop.
- Forewarning the outbreak of the dreaded inflorescence blight disease in west coast region of India is of utmost importance for its effective management.
- Proper training of cashew trees from early stages and timely pruning of side and criss-cross branches may ensure that microclimate is not congenial for disease build-up of inflorescence blight disease.

Mitigation Measures to Reduce Emission of CO_2/GHG and Carbon Sequestration Potential of Crops

Cashew has dense green leaves with good photosynthetic capacity and can also be grown in high-density planting system. Cashew is ideal crop for carbon sequestration. It can be grown in vast degraded/wasteland existing in cashew-growing regions. Other mitigation measures are as follows:

- Resource conservation techniques
- Organic farming.

FUTURE STRATEGIES FOR RESEARCH

- Soil and water conservation measures including mulching, supplemental irrigation, drip irrigation and fertigation during fruit development stages needs to be evaluated and developed to reduce loss of yield due to drought under climate change scenarios.
- Carbon sequestration potential of cashew plantations needs to be evaluated for mitigation of climate change.
- Quantification of relationship between climatic parameters and incidence of pests, particularly tea mosquito bug (TMB) and inflorescence blight at multiple locations, where the crop is cultivated.
- Development of suitable contingency measures for adaptation to aberrant and extreme climatic events.

CONCLUSION

The yield of rainfed cashew crop is largely dependent on climatic and weather conditions. The climate strongly influences the critical phenological development of reproductive phase and outbreak of important pests and diseases. Prolonged rains beyond October delays bud break and flowering in cashew. Heavy rains during harvesting period affect production and quality of cashew nuts. Late varieties with more heat units requirement yield more than early varieties requiring lesser heat units. In absence of moisture stress, late and medium varieties out-perform early varieties in terms of nut yield and kernel weight as well. Cloudy conditions, high relative humidity and heavy dewfall are conducive for outbreak of pests and diseases. The crop is sensitive to salty seawater and extreme climatic events of rainfall, drought and high wind velocity during reproductive phase. The perennial cashew crop has potential for carbon sequestering for mitigation of climate change. Little information is available on effect of elevated CO_2 and rise in temperature under future climate change scenarios.

References

Pathummal, Beevi, S. 2004. *Annual Report,* ICAR Adhoc Scheme on Forewarning Tea Mosquito Bug (*Helopeltis antonii*) in Cashew, pp. 82-85.

Prasada Rao, G.S.L.H.V. 2002. Climate and Cashew AICRP on Agro meteorology, Department of Agricultural Meteorology, College of Horticulture, Kerala Agricultural University, pp. 1-100.

8

Climatic Influence on Growth and Productivity of Black Pepper

K. S. Krishnamurthy, K. Kandiannan, B. Chempakam
Utpala Parthasarathy and V. A. Parthasarathy

HUMAN intervention has led to increased emissions of greenhouse agents (CO_2, NO_2, water vapor and methane) in the atmosphere tapping heat through ghgreenhouse effect far in excess than required, which is responsible for global warming and climate change. In the last 100 years, the global temperature has increased by 0.4-0.8 °C. The UNEP (United Nations Environment Programme) and WMO (World Meteorological Organisation) established the Intergovernmental Panel on Climate Change (IPCC). The IPCC Report (2001) stated that the carbon dioxide emitted mainly by the burning of fossil fuels and the emission of methane, nitrous oxide and other greenhouse gases thicken the blanket of green house gases over the earth's atmosphere, upsetting the natural flow of energy from the sun. Deforestation accelerates global warming by reducing the absorption of carbon dioxide in atmosphere. Based on current emission levels, it is believed that the global temperature will rise by between 1 °C and 3.5 °C between now and the year 2100. Global climate change will definitely affect agriculture. Anticipated changes in global rainfall and temperature patterns together with the increase in atmospheric CO_2 will affect the production of crops throughout the world.

IMPACT ANALYSIS ON BLACK PEPPER

Black pepper is sensitive to climatic requirements of temperature, humidity and rainfall. Before impact analysis of climate change and global warming on black pepper productivity, it will be pertinent to discuss its climatic requirements briefly.

Climatic Requirements

Temperature and relative humidity: Black pepper (*Piper nigrum* L.) is a plant of humid tropics requiring adequate rainfall and humidity. The crop tolerates a temperature between

10 and 40 °C. The optimum temperature range is 23–32 °C with an average of 28 °C. Optimum soil temperature for root growth is 26–28 °C (De waard, 1969; Wahid and Sitepu, 1987). It successfully grows between 20^0 North and South latitude and from sea level up to 1500 m (Radhakrishnan *et al.*, 2002). A relative humidity of 60–95% is optimum at various stages of growth. Tropical temperature and high relative humidity with little variation in day length throughout the year are most favourable for the crop.

Rainfall: Total rainfall and its distribution play most important role in black pepper cultivation and productivity. Annual rainfall of 2000–3000 mm with uniform distribution is ideal. Rainfall of 70 mm received in 20 days during may-June is sufficient for triggering flowering process in plants, but once process is set off there should be continuous shower until fruit ripening. Any dry spell even for a few days, within this critical period of 16 weeks (flowering to fruit ripening) will result in low yield (Pillay *et al.*, 1988). In India, black pepper growing areas receive 1,500 mm or more than 4,000 mm rainfall. Rainfall after stress induces profuse flowering (Pillay *et al.*, 1988).

Black pepper yield is significantly correlated with rainfall received (>100 mm to attain field capacity moisture in soil) during first half of May (r=0.75), and that received during second half of June (r=0.90), if preceded by rainfall in the first half of May. Total annual growth of fruiting branches are required to be maximized (82.43%) in June-July, coinciding with peak period of monsoon for higher productivity of the vine (Pradeepkumar *et al.*, 1999). High dry matter accumulation was observed in branches just before shoot elongation and flowering during April-May (Mathai, 1983). The late commencement of south west monsoon causes a delay in flower initiation of black pepper. Pollination and development of berries are affected due to the failure of adequate post-blossom rain (Remold, 1997). The data from Pampadumpara showed that good rainfall during March had positive influence on productivity. It was also noted at one of the farmers fields in Coorg, that good rainfall during March lead to increased productivity. The data also suggested that heavy rains during April followed by very low rains in May may hamper productivity.

Prolonged spell of drought or heavy rains or sharp and sudden alteration of both during advanced stage of berry development could lead to spike shedding. Intensive shedding occurs during years in which heavy north east monsoon showers are received after a spell of dry period after south-west monsoon (Sukumarapillay *et al.*, 1977).

Elevated CO_2 and Temperature

Information on the effect of elevated CO_2 under climate change and global warming scenarios on black pepper productivity is not available. Effects of temperature and rainfall crucial for black pepper are discussed here.

Growth and Productivity: Based on HadCM3 RCM predictions for 2020 and 2050 for the grid covering Calicut for A2a scenario, the increase in temperature (maximum) for Calicut is about 1.7° for 2020 and 2.7° for 2050. Similarly, increase in temperature (minimum) is about 0.5 and 1.5°, respectively for 2020 and 2050. Therefore, an experiment was set up in black pepper with 10 varieties in growth chamber with set temperatures of 2.7 and 2° increase (over 20 years average) in day and night temperatures, respectively. Growth parameters (plant height and leaf area at regular intervals) and diurnal variation in photosynthetic rate were measured and compared with those of present ambient conditions.

Growth parameters: The leaf area and plant height of different black pepper varieties maintained in growth chamber for four months at ambient as well as at increased temperatures as mentioned above were measured. The variation up to 10% was observed for both leaf area as well as plant height after 4 months of growth at elevated temperatures. Most of the varieties showed a small reduction in leaf area at elevated temperatures. Varietal variation was noticed for both plant height and leaf area.

Diurnal variation in photosynthesis: Diurnal variation in photosynthesis varied with the varieties. Some varieties such as Panniyur 1 showed decreased photosynthetic rate at elevated temperatures while Sreekara and Subhakara maintained similar levels of photosynthesis both at ambient as well as at elevated temperatures indicating that elevated temperatures of 2-3 degrees may reduce photosynthesis in a few varieties, though most of the varieties did not show a drastic decrease.

Recent Productivity Trends in Relation to Rainfall

Black pepper is mainly grown as a rainfed crop in homestead gardens. Recently, its productivity is showing a decreasing trend, which could be the result of increase in temperature and reduction in rainfall (Table 1). The data on rainfall, Temperature (maximum) and Temperature (minimum) were collected from black pepper-growing region of the country for the past two decades (1984–2004). The black pepper productivity data was also collected for the corresponding period from these regions (Table 8.1).

Table 8.1. Recent trends in productivity and climatic parameters in black pepper growing regions of India.

Place	*Rainfall*	*Temperature (maximum)*	*Temeperature (minimum)*	*Productivity*
		Kerala		
Ambalavayal (Wynad Dist.)	Decreasing	Increasing	Increasing	Decreasing
Pampadumpara (Idukki Dist.)	Decreasing	–	Increasing	Increasing
Calicut	Decreasing	–	–	Decreasing
Panniyur (CannanoreDist.)	Decreasing	Increasing	Increasing	Decreasing
Trichur	Dccreasing	Decreasing	No trend	Decreasing
TAMILNADU				
Valparai (Coimbatore Dist.)	No trend	No trend	Decreasing	No change
Nilgiris	Increasing	Increasing	No trend	Decreasing
		Karanataka		
Coorg	Decreasing	–	–	Decreasing

Productivity and Climatic Parameters: The correlation between black pepper yield and climate parameters was worked out. The climate and black pepper production data for the past two decades (1984–2004) from pepper growing areas of the country were used for the purpose. In Ambalavayal (Wynad), December rainfall showed negative correlation with productivity. Temperature (minimum) during March and January showed significant positive correlation with productivity while Temperature (maximum) did not show significant correlation (Table 8.2).

Table 8.2. The correlation coefficients of productivity with rainfall and temperature during various months.

Month	*Temperature (maximum)*	*Temperature (minimum)*	*Rainfall*
January	0.15	0.46*	-0.04
February	0.14	0.38	0.31
March	-0.07	0.58*	0.23
April	-0.23	0.23	-0.01
May	0.25	0.24	-0.31
June	0.12	0.23	-0.07
July	0.01	0.25	0.23
August	0.10	0.18	0.21
September	0.10	0.17	0.01
October	-0.04	0.12	-0.12
November	0.02	0.21	0.00
December	0.18	0.24	-0.41

In Nilgiris, rainfall during December and January had negative correlation with black pepper productivity. Number of rainy days during May had positive influence while that of January was negatively correlated. Temperature (minimum) showed significant positive correlation with pepper productivity during March, April, July, August and October. In Panniyur (Cannanore District) also, rainfall and Temperature (maximum) during December showed negative correlation with productivity and Temperature (minimum) from August to December showed significant negative correlation with productivity while in Trichur, Temperature (minimum) during January and Temperature (maximum) during May and July had significant negative influence on productivity. In Pampadumpara (Idukki District), October rainfall was positively correlated while June, July and September rainfall had negative influence on productivity whereas in Valparai (Coimbatore District), none of the climatic parameters showed significant correlation with productivity.

The correlation between climatic parameters and productivity shows in general that in plains (Trichur and Kannur) Temperature (minimum) and Temperature (maximum) had negative influence while in higher elevations (Wynad, Nilgiris, Pampadumpara) Temperature (minimum) had positive influence on productivity. Temperature (maximum) does not seem to have much influence on productivity in higher elevations. This implies that climate change especially the increase in minimum temperature could increase productivity in higher elevations while the same could reduce the productivity in plains. Increased rainfall during December and January tend to decrease productivity while April and May rainfall increase the productivity.

Correlation between pepper productivity and climatic parameters based on climatic data for 11 years showed that maximum temperature and number of sunshine hours in the first fortnight of March had positive impact on productivity, while mean relative humidity of the July first fortnight, number of sunshine hours received during the February first fortnight and April second fortnight, and mean maximum temperature during the June second fortnight had negative influence on productivity (Pradeepkumar *et al.*, 1999).

The regression equation obtained based on 10 years climatic data in eight plantations showed that Temperature (maximum) and Temperature (minimum) may have more influence on productivity than rainfall or rainy days. Temperature (maximum) showed negative influence which indicates that climate change in terms of increase in Temperature (maximum) may have negative influence on productivity. Temperature (minimum) showed positive influence in 50 % of the plantations.

Influence of rainfall deficit on black pepper productivity was also studied. Rainfall during 1987 and 2002 (declared as drought years by the government) was less than normal (mean of 1984–2004) for all the black pepper-growing areas during those years. The black pepper productivity also was below normal (mean of 1984–2004) for those years, indicating the negative influence on rainfall deficit on black pepper productivity. The rainfall deficit varied from 6.8–19.3%, while the yield reduction varied from 26.6–57.8% over 20 years mean during 1987. Similarly, during 2002, rainfall deficit varied from 28.46–43.43% while the yield reduction varied from 3.9–39%. This clearly shows that rainfall deficit has a great negative impact on the productivity of black pepper.

Pepper productivity is generally higher in higher elevations such as Wynad and Idukki (800–1,000 m above msl) compared to low elevations such as Cannanore and Trichur (0 to 200 m above msl). The temperature (20 years mean of both Temperature (maximum) and Temperature (minimum)) of both Idukki and Wynad is about 5–7° lesser than that of Cannanore or Trichur. Relatively cool climate of these regions may have positive influence on productivity. Mean rainfall of 20 years shows that rainfall is higher in Trichur/Cannanore compared to Wynad/Idukki (Table 8.3).

Table 8.3. Mean productivity, temperatures and rainfall in important black pepper growing regions of India.

Region	*Mean Temperature (maximum) (°C)*	*Mean Temperature (minimum) (°C)*	*Mean Yield (kg/ha)*	*Mean Rainfall (mm)*
Wynad	27.3	17.6 (9.7)	402	1972
Idukki	27.5	15.6 (11.9)	327	2010
Cannanore	33.1	22.6 (10.5)	241	3433
Trichur	32.1	23.4 (8.7)	239	2753

Quality

To study the influence of climate (mainly temperature) on quality parameters of black pepper, berry samples were collected from about 95 locations covering major black pepper-growing regions of Kerala, Karnataka and Tamil Nadu. There is a difference of 2-3° in mean maximum and minimum temperatures among these regions (if we consider plains and higher elevations separately). Rainfall of most these regions is within the range suggested for black pepper cultivation. So, comparison of quality parameters among these regions may reveal the future climate change effects on quality as predicted temperature increase is about 2-3° for the coming century in black pepper-growing areas (Table 8.4). The results revealed that though there is variation in oil, oleoresin and piperine content of samples from different places, temperature does not seem to play major role. For example, the oil content of samples from Kasaragod, Calicut, Kottayam, Aleppey and Trivandrum, all is around 3.0%. Similarly, piperine content for these places is around 4.0%. There was some difference in oil components from these places. The mean maximum and minimum temperature differences among these places is around 1-2°. But samples from higher elevation (Wynad and Idukki) showed slightly lower oil and piperine content. Differences in oil components (mainly beta caryophyllene, the major component) was also noticed. Karnataka and Tamil Nadu samples had lesser oil and piperine content than Kerala samples and mostly, these are under higher elevation (Table 8.4).

Table 8.4. Correlation between elevation and quality

Parameter	*Panniyur 1*	*Karimunda*	*Panniyur1+ Karimunda*
Oil	- 0.11	-0.02	NS
Oleoresin	- 0.05	-0.28	NS
Piperine	_	-0.06	NS
α – Pinene	0.28*	0.21	0.25*
Limonene	0.30*	0.28	0.27*
β – aryophyllene	-0.53**	-0.45	-0.48**

The data shows that oil, oleoresin and piperine tend to decrease with elevation in both Karimunda and Panniyur-1 varieties (though non-significant). The main essential oil component β – caryophyllene also showed significant decrease with increase in elevation, while other two components á-pinene and limonene showed significant increase with increase in elevation.

Pests of Target Crop

The average annual yield loss in black pepper due to drought was 4%, due to foot rot was 9%, due to pollu beetle 13% and due to fungal pollu was 5% as reported by Prabhakaran (1997) in a study conducted in Kannur district of Kerala during 1989-92. The *Phytophthora palmivora* infection of black pepper in an arecanut-black pepper mixed cropping system was positively correlated with rainfall, number of rainy days and relative humidity (Ramachandran *et al.*, 1988; Ravindran and Johny, 2002; Jayasekhar and Muthuswamy, 1999) and was negatively correlated with temperature and sunshine hours (Ramachandran *et al.*, 1988). Climatic parameters are major determinants of *Phytophthora* leaf infection in black pepper (Table 8.5). A daily rainfall of 15.8-23 mm, RH of 81-99%, temperature range of 22.7–29.6 °C and sunshine hours of 2.8-3.5/day achieved during the peak monsoon period of year are conducive for maximum development of disease (Anandaraj and Sarma, 1994). A temperature range of 18.6-22.5 °C along with a relative humidity of 95% and well distributed rainfall are conducive for maximum development of foot rot in Karnataka.

The maximum disease incidence was noticed during August and was correlated positively with relative humidity and negatively with maximum temperature prevailed during the period (Shamarao and Sidharamiah, 2002). The increase of soil temperature from 24.5–29.63 °C reduced *P. capsici* population from 89.76–25.76% (Hegde and Anahosur, 1998). Outbreaks of *P. nigrum* root rot were enhanced by long periods of cloudy, humid weather with occasional rainfall and by deep shading (Matsuda *et al.*, 1994).

Berry damage caused by pollu beetle (*Lanka ramakrishnai*) was highest in the plains, and it was very low at higher elevations (300–900 m above MSL) and absent at > 900m MSL. The damage was also higher in shaded than in open positions (Kumar and Nair, 1987). Extensive feeding by adults and mating were observed during pre-monsoon showers in April-May. There was a significant correlation between adult population and temperature; high temperatures adversely affected the pest population. Rainfall was favourable to population build-up (Kumar and Nair, 1985). Leaf gall thrips population showed a significant positive correlation with rainfall, minimum temperature and relative humidity and a significant negative correlation with maximum temperature. The incidence of all these insect pests in various locations was affected by altitude. The maximum incidence of pollu beetle was observed in locations below 300 m MSL, while that of leaf gall thrips and scale insects was significantly higher at higher altitudes (above 750 m MSL). Leaf gall thrips incidence was high during the monsoon season (June–September) and low during the post-monsoon and summer seasons (January–April), while the incidence of scale insects was higher during the post-monsoon and summer seasons (January—April).

Table 8.5. Climatic factors (means) during increasing and decreasing phases of *Phytophthora* leaf infection in black pepper

	Year	*No. of observations*	*Temperature (°C)*		*Relative humidity (%)*		*Rainfall (mm)*		
			Min.	*Max.*	*Max.*	*Min.*	*Per day*	*Days/week receiving 10 mm and above*	*Sunshine (hours/day)*
Increasing phase	1984	6	26.0	29.0	82.0	94.4	15.8	2.80	3.5
	1985	9	22.7	27.3	83.8	99.2	22.2	4.81	2.8
	1986	5	24.0	29.6	81.0	97.9	23.0	4.40	3.2
Decreasing phase	1984	5	26.3	28.6	82.4	95.6	8.4	1.80	4.8
	1985	8	23.6	28.7	76.4	97.9	8.4	2.20	3.5
	1986	10	24.7	30.8	70.9	88.9	10.0	1.80	6.8

Source: Ramachandran *et al.*, 1988

OTHER INDIRECT EFFECT OF CLIMATE CHANGE AND GLOBAL WARMING

Future geographical distribution: Future prediction shows an increase in temperature of 1.5-2.5 degrees by 2050 in the country in general. Keeping this change in temperature in view, land suitability changes was worked out for black pepper using the eco-crop model of DIVA GIS. The model predicts suitability of Western Ghats region where black pepper is predominantly grown from excellent to marginal while it predicts a very low change in suitability level for North Eastern Himalayas. The DIVA-GIS was also used to predict the possible distribution of Piper species in other parts of India. Based on these prediction distribution maps, it is expected that most areas of north-east India, coastal Andhra Pradesh, Orissa and West Bengal are also likely to have many piper species, while the chances of distribution of Piper species in other parts of the country are remote. The distribution of Piper species at different altitudes was also studied and mapped using DsIVA-GIS. Interestingly, all the species occur in the Silent Valley National Park in the Western Ghats.

Aberrant Weather and Extreme Events

- Severe spike shedding during berry development stage, in years with heavy showers of north-east monsoon after prolonged dry spell of summers.
- Dry spell of even a few days during critical 16 week period of monsoon, coinciding with flowering to fruiting reduces yield drastically.

Adaptation Measures for Climate Change and Global Warming

- Mulching the pepper basins using locally available material/coconut husk conserves soil moisture.
- Use of drought tolerant varieties such as Panniyur 6, Panniyur 7, accessions 1495, and 931 etc.
- Irrigation at an IW/CPE ratio of 0.25 to enhance pepper yields.
- Irrigating the pepper basins at least once during March and April @ 30–40 litres per vine.

Mitigation Measures to Reduce Emission of CO2/GHG and Carbon Sequestration Potential of Crops

Carbon sequestration potential of perennial black pepper vine has not been estimated. Other mitigation measures are as follows:

- Resource conservation techniques
- Organic farming.

FUTURE STRATEGIES FOR RESEARCH

- Development of relatively drought tolerant varieties capable of withstanding reduced precipitation and uneven distribution of monsoon showers.
- Detailed studies to identify effect of increased temperature and CO_2 concentration on growth, yield and quality.
- Formulating adaptation strategies not only for black pepper but also for all other spice crops.
- Systematic studies on extreme events such as floods or drought during different growth stages of various spices.
- Site suitability studies under situations of increased temperature and reduced rainfall.
- Studies on impact of climate change on pest incidence, their population dynamics and management.
- Technology development for artificial rain making guns in dry spells during critical flowering and fruiting.
- Development of forewarning systems for aberrations in monsoon to effect remedial measures in advance.

CONCLUSION

Black pepper is mainly grown as a rainfed crop in homestead gardens. Productivity highly depends on timely and well-distributed annual monsoon of 2000–3000 mm. Recent trends in black pepper productivity showed a decreasing trend with concurrent decrease in rainfall and increase in temperature in major black pepper-growing areas of India. Increased rainfall during March-April increases black pepper productivity, while December and January rainfall has negative impact. Minimum temperature is positively correlated in higher elevations (900–1000m above MSL), while both Temperature (maximum) and Temperature (minimum) have negative impact on productivity in plains (50–300 m above MSL). Increase in temperature due to climate change may increase productivity in higher elevations such as Wynad, while it may have negative impact in plains. Warming and increased temperature (1-2 °C) did not influence quality parameters such as oil, oleoresin and piperine content. Main quality component of beta caryophyllene was positively correlated with increase in temperature. The growth components such as leaf area, plant height and photosynthesis was about 5–10% lesser in black pepper varieties maintained at an elevated temperature of about 2 °C compared to that of ambient temperature. The land suitability changes worked out for black pepper using the eco-crop model of DIVA GIS predicts suitability of Western ghat region where black pepper is predominantly grown as highly suitable at present. However, it may change to marginal suitability with rise of 1.5-2 °C in temperature. The model predicts a slight change in suitability level for North Eastern Himalayas with 1.5-2 °C increase in temperature.

References

Anandaraj, M. and Sarma, Y.R. 1994. Biological control of Black pepper diseases. *Indian Cocoa, Arecanut and Spices Journal* **18:**22-23.

De waard, P.W.F. 1969. Foliar diagnosis nutrition and yield stability of black pepper (*Piper nigrum* L.) in Sarawak. Communication No. 58, Department of Agrcultural Research, Koninklijk Instituut Voor de Tropen, Amsterdam.

Hegde, S.G. and Anahosur, K.H. 1998. Integrated management of foot rot of black pepper. *Karnataka Journal of Agricultural Sciences* **11:**78-82.

Jayasekhar, M. and Muthuswamy, M. 1999. Influence of weather parameters on the incidence of foot rot of black pepper. *Madras Agricultural Journal* **86:**344-46.

Kumar, T.P. and Nair, M.R.G.K. 1985. Seasonal fluctuation of population and habits of the adults of *Longitarsus nigripennis* Mots. on black pepper. *Journal of Plantation-Crops* **13:**138-41.

Kumar, T.P. and Nair, M.R.G.K. 1987. Effect of some planting conditions on infestation of black pepper by *Longitarsus nigripennis* Mots. *Indian Cocoa, Arecanut and Spices Journal* **10:**83-84.

Mathai, C.K. 1983. Growth and Yield analysis in black pepper varieties (Piper nigrum L.) under different light conditions. *Mysore Journal of Agricultural Sciences* **18:**317.

Matsuda, A., Hamada, M. and Gonzalez, J.L. 1994. Outbreak and control of foot rot on black peppers in Dominican Republic. *Agrochemicals Japan* **64:**23-26.

Pillay, V.S., Sasikumar, S. and Ibrahim, K.K. 1988. Effect of rainfall pattern on the yield of black pepper. *Agrometeorology of Plantation Crops* 152-59.

Prabhakaran, P.V. 1997. Quantitative determination of loss in yield of black pepper (*Piper nigrum* L.) in Kannur district (Kerala, India). *J. Spices and Aromatic Crops* **6:**31-36.

Pradeepkumar, T., Vasanthakumar, Aipe, K.C., Kumaran K, Susamma P George, Manmohandas TP and Anith KN. 1999. Studies on yielding behaviour of black pepper CV Panniyur-I. *Indian Journal of Arecanut Spices and Medicinal plants* **1:**88-90.

Radhakrishnan, V.V., Madhusoodanan, K.J., Kuruvilla, K.M. and Vadivel, V. 2002. Production technology for black pepper. *Indian Journal of Arecanut, Spices and Medicinal plants* **4:**76-80.

Ramachandran, N., Sarma, Y.R. and Anandaraj, M. 1998. Effect of climatic factors on Phytophthora leaf infection in black pepper grown in arecanut- black pepper mixed cropping system. *Journal of Plantation Crops* **16:**110-18.

Ravindran, P.N. and Kallupurackal Johny A. 2002. Study of black pepper research under AICRP on spices. *Journal of spices and Aromatic crops* **11:**122-24.

Remold, M. 1997. Pepper prospects in India during 1997-1998 season. *Indian Cocoa, Arecanut and Spices Journal* **21:**1-5.

Shamarao, J. and Sidharamiah, A.L. 2002. Influence of weather forecasters on the epidemology of foot rot of black pepper in Karnataka. *Indian Journal of Agricultural Research* **36:**49-52.

Sukumarapillay, V., Sasikumaran, S. and Venugopalan, Nambiar, P.K. 1977. A note on preliminary observation of spike shedding in pepper. *Arecanut and Spices Bulletin* **8:**93-94.

Wahid, P. and Sitepu, D. 1987. Current status and future prospects of pepper development in Indonesia. Food and Agricultural Organisation, Regional office for Asia and Pacific, Bangkok.

9

Impact of Climate Change on Citrus in India

A. D. Huchche, P. Panigrahi and V. J. Shivankar

INDIA stands sixth in the world with 5896.0 thousand tonnes of yearly citrus production (FAO, 2006). Collectively, citrus fruits rank third in area and production (after banana and mango) and account for 9.28 and 5.21% in total area and production of fruits, respectively in India. Indian citrus industry is dominated by mandarins followed by limes/lemons (Anonymus 2007). Citrus is an evergreen subtropical crop, which has been acclimatized in tropical environs too. The most limiting factor restricting its geographical distribution is low temperatures. Frost and freezing damages the fruits and if persists long enough may kill the trees. Even at milder, non-damaging range, temperatures present major limitations for vegetative growth as well as for fruit development and maturation.

There is a considerable variation in productivity depending on orchard management and weather conditions prevailing at critical growth stages. Average productivity of citrus in Andhra Pradesh is about 14 tonnes/ha, whereas in some areas in Vidarbha (Maharashtra) a yield of 19–25 tonnes/ha is not uncommon. This indicates that there is ample scope for growth of citrus industry in future. There has also been a shift in area under sweet orange cultivation from western Maharashtra to Marathwada region to remunerative cash crops like sugarcane and grapes over the past 30 years. The shift was primarily driven by increased incidence of citrus greening disease and other insect-pests due to high humidity coupled with warm temperature.

Citrus productivity is projected to be16 tonnes/ha by the year 2025 to increase the per capita availability to 6.88 kg/year as compared to 3.10 kg/year at present. However, challenges imposed by climate change and global warming and associated ill effects like untimely or delayed rains, scanty rains, inordinately longer dry spells, hailstorms, extended winters and changes in pest complex *etc.* needs to be assessed accurately to evolve adaptation measures and research strategies.

IMPACT ANALYSIS ON CITRUS

Elevated CO_2 and Temperature

Citrus production is likely to be benefited by elevated CO_2 and rise in temperature due to increased photosynthesis, growth, flowering, fruit retention, fruit weight and rapid colouration, provided other factors do not change.

Productivity

***Elevated CO_2*:** The response to elevated CO_2 of 3-yearold fruiting Valencia orange scions [*Citrus sinensis* (L.) Osbeck] on citrange rootstock [*C. sinensis* × *Poncirus trifoliata* (L.) Raf.] was studied over a 12-month period under controlled environmental conditions (Idso and Kimball, 1997; Downton *et al.*, 2007). The CO_2 enrichment to approximately 800 μbar CO_2 which commenced just prior to anthesis shortened the period of fruit let abscission. Trees enriched with 800 μbar CO_2 retained 70% more fruits, which at harvesting were not significantly smaller in diameter or lower in fresh weight than fruits from control trees grown at approximately 400 μbar CO_2. The progression of fruit colouration was more rapid for CO_2 enriched trees. Changes in dry matter partitioning resulted in greater fruit yield (58% increase in dry weight) with CO_2 enrichment.

Joseph *et al.* (2002) observed that both mature (old) and expanding (new) leaves of two-year old trees of 'Ambersweet' orange, a hybrid of 'Clementine' tangerine (*Citrus reticulata* Blanco) and 'Orlando' tangelo (*C. paradisi* Macf. × *C. reticulata*) grown for 29 months under elevated CO_2 (720 ìmol mol^{-1}) and at temperatures of 1.5 and 6.0 °C above ambient temperature had higher photosynthetic rates, lower transpiration and conductance, and higher water-use efficiency (WUE), compared to those grown under ambient CO_2 (360 ìmol mol^{-1}). Although leaf WUE was reduced by high temperature, elevated CO_2 compensated for adverse effect of high temperature on leaf WUE. Activity and protein concentration of rubisco were down-regulated in both new and old leaves at elevated CO_2. The photosynthetic acclimation of Ambersweet orange leaves at elevated CO_2 allowed an optimization of nitrogen use by reallocation/redistribution of nitrogen resources away from rubisco. Thus, in the absence of other environmental stresses, citrus photosynthesis would perform well under rising atmospheric CO_2 and temperature as predicted for this century.

Temperature: Little growth occurs in all citrus tree organs below 13 °C. One of the prominent effects of climate change having direct bearing on the economy of citrus orchardists is the flowering and fruiting. The effects of temperature on flower induction in sweet orange were studied over a range of shoot and root temperatures representing the field conditions in commercial growing regions of Australia (Moss, 1969). Following low temperature inductive conditions (5 weeks at 15/10 °C), profuse flowering occurred at 24/19 °C; flowering was only slightly less at this air temperature with a root temperature of 30 °C, and at an air temperature of 30/25 °C. At 36/31 °C a few flowers were formed and these abscised before opening. However, when root temperature was maintained at 22 °C with this shoot temperature, a few flowers were formed. In another experiment, a root temperature of 30 °C and shoot temperature of 24/19 °C reduced the number of flowers that formed. Only a small part of the effect of high temperature in inhibiting flower initiation was due to root temperature. It was concluded that flowering could be induced in field during summer by low night temperature and root temperatures would probably have little influence on floral development.

The optimum temperatures for resumption of growth in citrus after winter rest, ranges from 13 to 37 °C. In central India, temperatures below 13 °C in January delay initiation of flowering in Nagpur mandarin. The month of March is another crucial period for the growth of tender fruitlets. Although minimum temperatures remain favourable for growth, but if maximum temperatures exceeds above 37 °C, tender fruitlets are seriously damaged. As a result, there is a heavy drop of small fruitlets causing drastic reduction in yield or crop failures at times. The month of May experiences highest maximum temperatures in central India. The average maximum temperatures reach 44-45 °C in May, causing slowing down of fruit growth and sometimes causing excessive fruit abscission.

In subtropical climates, colour break occurs when night temperatures fall below 12.8 °C. Low soil temperatures (<12 °C) promotes rapid colour development. The rate of colour development after colour break, is influenced by the extent of the subsequent temperature drop, the duration of low temperature and physiological state of cultivar concerned. Internal acid levels on the other hand, do not drop as quickly with low temperatures as they do with relatively high night temperatures. Fruits that mature internally with night time temperatures below 13 °C thus do not have a colour problem, but may have higher levels of acidity than fruit maturing internally at night time temperatures above 13 °C, which do have a problem (Grierson *et al.*, 1986).

Quality

Information on effect of elevated CO_2 and rise in temperature on quality of citrus produced is not available.

Pests of Target Crop

Pests: Data on population dynamics of major insect pests and their predominant bioagents were recorded and correlated with weather parameters during 1998–2005. *Hasta* season (October-November) was favourable for parasitoids of leaf miner, whereas *Ambia* (February-March) season was favourable for predators in general and for parasitoids in particular, e.g., *Tamarixia radiate* of citrus psylla and *Encarsia* spp. of blackfly. Predatory and parasitic activity was found to be pronounced in Vidarbha and Punjab, respectively.

Psylla incidence is negatively correlated with RH and has direct relationship with temperature. The pest multiplies enormously during the long dry spell in July. The incidence of mites aggravates during summer in the absence of humidity. The blackfly parasitoid *Encarsia* spp. and citrus psylla survival is directly proportional to higher temperature and inversely proportional to RH (Anonymous 2005).

The citrus leaf miner incidence (%) is positively correlated with maximum and minimum temperature, rainfall, wind velocity and negatively correlated with relative humidity, vapour pressure and sunshine hours (Table 9.1). The parasitism (%) is correlated negatively with maximum temperature and rainfall, whereas it is positively correlated with relative humidity, vapour pressure and sunshine hours.

Table 9.1. Correlation between incidence of leaf miner, eulophid parasitism, predatory population and weather parameters

Weather parameter	*Citrus leaf miner incidence (%)*	*Parasitism by eulophids (%)*	*Predatory population (M. boninensis and M. sexmaculatus)*
Maximum temperature	0.252	-0.397	-0.184
Minimum temperature	0.180	-0.669*	-0.680*
Relative humidity	-0.456	0.170	-0.517
Rainfall	0.252	-0.099	0.412
Vapour pressure	-0.391	0.046	-0.876**
Wind velocity	0.509	-0.913**	-0.018
Sunshine hours	-0.002	0.414	0.490

*P = 0.01 ; **P = 0.05

The Mediterranean fruit fly, *Ceratitis capitata* (Wiedmann), is a quarantine pest in many countries where it is not endemic. The accepted technology for controlling this insect on citrus fruits is a cold quarantine procedure where the fruit is held at 1-2 °C for a period of 16–21 days (Lurie, 2006). This method is effective but since citrus is a subtropical fruit, this extended length of time at low temperature can cause chilling injury. A high temperature thermal treatment is more efficient method of controlling the fruit fly. By adding an element of controlled atmosphere to high temperature, the time of exposure required can be shortened, or the temperature of exposure can be lowered. The effect of temperatures 40, 42 and 44 °C on different developmental stages of fruit fly was examined and it was found that most resistant stage to high temperature was the third instar. In addition, it was found that even the third instar larvae that survived were impaired in their ability to develop into mature fruit flies.

***Disease incidence*:** In citrus, environmental factors play an important role in disease development, severity, dispersal, and survival of *Phytophthora* in causing epidemics. The *P. palmivora* thrives best at 25–28 °C in Maharashtra (Das, 2008). Distribution and activity of *P. palmivora* was related with cooler and humid region of Maharashtra and Coorg where temperature remained below 28 °C and rainfall was above 1500 mm, while *P. nicotianae* was prevalent in warmer and humid climate of Vidarbha, Ahmednagar and Kodur at 30–32 °C temperature and rainfall below 1150–1300 mm. In citrus nurseries of Vidarbha region, *P. citrophthora* was active during winter season at 20–24 °C temperature and 60-70% RH and ceased its activity with the increase in temperature while population of *P. nicotianae* and *P. palmivora* increases. The soil temperature in central India remains between 20 and 29 °C throughout the year. Thus *Phytophthora* remained active throughout the year in nurseries and in irrigated orchards. The problem of *Phytophthora* has been severe during the last decade.

Sporadic epidemics of *Phytophthora* brown rot occur in citrus orchards in Florida and are caused primarily by *P. palmivora* and secondarily by *P. nicotianae*. With *P. palmivora*, fruit wetness durations of 3 hours or more resulted in maximum infection at favourable temperatures. The optimum temperature range for fruit infection and brown rot development was 27–30 °C. No brown rot developed at 22 °C or less. The optimum temperature for sporulation in culture and on fruit surface was 24 °C, with sporangium production decreasing rapidly at higher or lower temperatures. Brown rot caused by *P. palmivora* can become epidemic in Florida from

July—October during the rainy season because of its high temperature and prolific sporulation on the fruit surface (Timmer and Zitko, 2000).

The citrus bacterial canker disease is favoured by temperatures between 20 and 30 °C and evenly distributed rains (Das, 2002). Citrus canker is severe in regions where temperature and rainfall ascend and descend simultaneously during the year. Therefore, the disease occurs in severe form in seasons and/or areas characterized by warm and humid weather conditions.

Climatic Trend Analysis in Citrus Growing Areas of Central India

The climatic factor of rainfall has an important bearing in rainfed citrus production influencing water stress, soil erosion, and disease and pest problem. The past trends were analyzed to indicate effect of climate change on rainfall pattern, water availability and nutrient losses in central India.

Rainfall: Maximum rainfall in central India occurs during June—September which averages to be 1000–1100 mm annually. However, over the past 10 years, there has been a substantial reduction in annual rainfall (700-800 mm). Due to dwindling annual rainfall (Fig.9.1) and expansion of area under citrus cultivation, the ground water table in central India is declining. Thus, during the years of deficit rainfall, the recharge of the ground water becomes less. Since a majority of orchards in Vidarbha is irrigated by open wells, the availability of water during peak summers is a limiting factor. This leads to excessive water and temperature stress damaging the plant health.

During the last three decades the rainfall pattern has changed dramatically. Late onset of monsoon followed by long dry spells and maximum precipitation occurring during July—August lead to maximum runoff and least percolation/recharge. Early rains (May), incessant rainfall

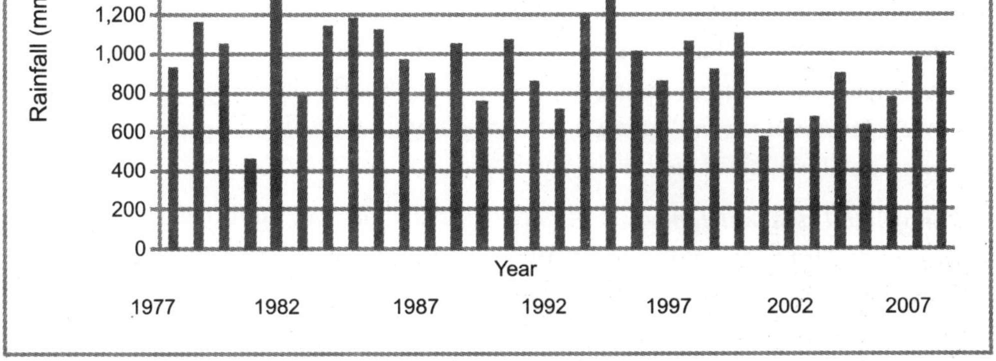

Fig. 9.1. Yearly average rainfall during 1977–2007

for more than 10 days (July 2007), long dry spells (July, 2008), scanty rains in August, 2008, early rise in temperature (around 40 °C) in the last week of March (2005–2008), late flush in spring is exposed to such high temperature. Absence of optimum RH (60–65%) during dry spells to retain the growth of the young developing fruits leads to heavy fruit drop, off-season hail storms in spring (1996, 2006) are some of the quotable examples regarding vagaries of climate.

***Water requirement and crisis in Vidarbha*:** Proper water management is essential for better fruit yield and quality in citrus. In general full grown (>5 years old) citrus trees require 1000-1500 mm water/annum to meet evapo-transpiration (ET) needs. Erratic and scanty rainfall disturbs the irrigation scheduling of citrus and create stress condition in root zone for the rainfed crop.

Table 9.2. Water requirement and yield changes of Nagpur mandarin during 2003-07 under drip irrigation

Year	*Water applied (litres/day/plant)*	*Yield (tonnes/ha)*
2003	72–230	18.6
2004	85–255	16.8
2005	80–250	17.5
2006	70–210	18.3
2007	75–200	15.6

Experiments conducted at NRC for Citrus, Nagpur, showed that volume of irrigation water varied from plant to plant for a mature bearing tree of Nagpur mandarin (Table 9.2). The minimum and maximum water were applied in December and May in response to lowest and highest evaporation, respectively. However, irrigation water quantity varied from year to year in the range of 10–15% due to changing environmental conditions without much change in yield.

The water crisis in Vidarbha region is summarized as under:

- Average total annual rainfall in the last 30 years = 1,028 mm
- Average total annual rainfall in the last 10 years = 772 mm
- Total annual rainfall deficit in the last 10 years = 256 mm
- Average annual rainfall during rainy season in last 30 years = 873 mm
- Average annual rainfall during rainy season in last 10 years = 657 mm
- Net rainfall deficit in the last 10 years = 216 mm

***Runoff, soil and nutrient loss*:** It was observed that the run-off generated in Nagpur mandarin orchards (3.2%s slope) on vertisols varied from 13.2–26.4% during 2003–07 (Table 9.3). It was only due to changes of rainfall characteristics and other climatic parameters. Due to changes in run-off pattern, the annual soil loss from the orchard also changes from 3.2–5.52 tonnes/ha.

Table 9.3. Rainfall, run-off and soil loss in Nagpur mandarin orchard during 2003-07

Year	*Rainfall(mm)*	*Run-off (% of rainfall)*	*Soil loss (tonnes/ha)*
2003	808	26.4	4.60
2004	666	22.5	3.20
2005	1034	13.2	5.52
2006	895	19.0	4.81
2007	1012	20.6	4.98

Keeping the loss of run-off, soil and nutrients in view, various *in situ* conservation measures were evaluated at NRC for Citrus, Nagpur, during 2003–07. The conservation measures tested were continuous bunding, continuous trenching and staggered trenching. Continuous trenching was found most suitable in conserving highest runoff (30–35%) and soil loss (25–30%) over control (without any conservation measures). Moreover, runoff harvesting through a farm pond of size 35m × 35m × 3m was successfully recycled to 5-year-old Nagpur mandarin plants of one hectare through drip irrigation and plastic mulch (Panigrahi *et al.,* 2007).

Aberrant Weather and Extreme Events

- Delay in monsoon, long dry spells between rains, untimely rains during water stress period, supra-optimal temperatures during flowering and fruit growth, hailstorms during spring, i.e., *Ambia* and *Mrig* crop harvesting are some of the most commonly encountered climate related problems experienced by the citrus growers over the past decade or so (Huchche, 2008).
- In central India, the temperatures below 13 °C in January delay initiation of flowering in Nagpur mandarin as observed during 1996, 1998, 2001, 2003 and 2006.
- The month of March is another crucial period for the growth of tender fruitlets. Although the minimum temperatures remain favourable for growth, maximum temperatures if exceeds above 37 °C as in the years 1994, 1999, 2001, 2003, 2004, 2006 and 2008 may cause serious damage to tender fruitlets.
- The month of May experiences highest maximum temperatures in central India. The average maximum temperatures reach 44-45 °C in May, causing slowing down of fruit growth and sometimes causing excessive fruit abscission, which was pronounced during 1998 and 2005.
- As a consequence of climate change, Indian citrus industry faces serious problems of insect-pests and diseases in central India. Epidemics of citrus black fly spanning over mid 70s, mid 90s in central India and Punjab, citrus psylla during 1960–62, 2000 and 2005–08 (monsoon), citrus leaf miner (1981, 1998, 2004, 2006) and occurrence of citrus rust mite during October-January 2007-08 over entire central India on all the commercial citrus fruits have been observed.

Adaptation Measures for Climate Change and Global Warming

- Adoption of microirrigation system for maintaining the plantation.
- Use of grass mulch (10–12 kg/basin) or black polyethylene mulch (100 micron) in plant basins throughout the growth period except monsoon.

- Use of Rangpur lime rootstock tolerant to drought, *Phytophthora* and soil salinity.
- Drainage should be provided in orchards along and across the slope.
- During dry spells supplementary irrigation should be provided for maintaining the growth of flowers/fruits. Use of growth regulators like GA_3 and 2,4-D in combination with foliar application of urea and other NPK mixtures has come in as a handy tool to manage fruit drop and increase fruit size.
- Fertilizer application through drip irrigation (fertigation) should be encouraged over conventional method of fertilizer application (broadcasting) in water stressed condition.
- Conservation of rainwater *in situ* through efficient conservation practices like continuous trenching in citrus on vertisols.
- Adoption of proper drainage system, i.e., bedding system in monsoon season to drain out excess rainfall and collection of this water through farm pond.
- Indiscriminate exploitation of groundwater must be avoided and groundwater recharging through rainwater harvesting by constructing check dams, percolation tanks and farm ponds in watershed scale and proper use of it for irrigation.

The NRC for Citrus, Nagpur, has devised following recommendations to overcome the ill effects of weather aberrations (Table 9.3).

Table 9.3. Effects of weather changes on crop growth in citrus and corrective measures

Type of weather change	*Adverse effects*	*Measures to be undertaken*
***Ambia* crop (January flowering)**		
Untimely winter rains during water stress period in December	• Water stress is broken, root system gets activated leading to vegetative growth.	• Foliar application of chlormequat chloride (CCC) 1500-2000 ppm of water after the rains
Sudden rise in temperature around 40 °C in second for tnight of March	• Heavy drop of fruitlets takes place	• Alleviation of water stress early January by adopting microirrigation with mulch
Hailstorm during fruit development period (February—April)	• Extensive damage of foliation, bark of trunk and stems, heavy drop of immature fruits, lesions on the leaves and fruits due to fungal infection	• Application of 2,4-D or GA_3 15 due to fungal infection.ppm along with urea 1.5% and carbendazim 1000 ppm to check further loss
Extremely high temperatures (45–47 °C) coupled with dry, hot winds in the month of May	• Excessive defoliation, fruit drop and cessation of root growth	• Light and frequent irrigations with mulching may be given. Foliar application of KNO_3 1.5% and planting of wind breaks on the borders
Delay in onset of monsoon of two-three weeks from the normal date	• Fruits growth is affected	• Mulching should be done either with grass or polythene sheet till the onset of monsoon. Life saving irrigations may be given to the existing crop and foliage should be saved from insect pests

(Contd.)

Type of weather change	*Adverse effects*	*Measures to be undertaken*
***Mrig* crop (June flowering)**		
Untimely rains in summer (May) during water stress period	• Water stress is broken, root system gets activated leading to vegetative growth	• Foliar application of chlormequat chloride CCC 1500–2000 ppm.
Dry spell during flower emergence and fruit set	• Flower initiation is affected	• Adoption of microirrigation system for ensuring continuous soil moisture. Also use KNO_3 @ 1.5% foliar spray at 10 days interval.
Delayed monsoon in late June to mid July resulting in late flowering (less degree days for fruit development)	• Fruit maturity period enters to late summer in April-May, where water scarcity is encountered	• Use fruit sizing foliar applications of monopotassium phosphate or potassium nitrate or diammonium phosphate @ 2% concentrationalong with either 2,4-D or GA_3 15 ppm at monthly interval till March (fruit maturity)
Dry spell condition after initial rains for 2-3 weeks	• Growth of fruitlets is affected. Incidence of insects like psylla becomes severe leading to blossom drop	• Rain water harvesting and using it as life saving irrigation during the dry spell.
		• Pesticide spraying should be adopted against insect pests and diseases
Incessant rains for more than a week during flowering	• Heavy drop of flowers and fruit setting is affected	• Application of GA_3 10 ppm + boron 0.3% for better fruit set & development.
Early withdrawal of monsoon i.e. by last week of August (late season drought)	• Water stress to the young fruitlets and mite infestation may occur	• Scheduling drip irrigation with mulch after the withdrawal of monsoon. • Acaricide application should be followed

Mitigation Measures to Reduce Emission of CO_2/GHG and Carbon Sequestration Potential of the Crop

Citrus plantations being perennial orchard have significant carbon sequestration potential, but the same has not been assessed and information is lacking. Other mitigation measures are as follows.

- Resource conservation techniques
- Organic farming

FUTURE STRATEGIES FOR RESEARCH

- Technology for inducing *Hasta bahar* (September-October flowering) needs to be standardised for different commercial citrus species.

- Canopy management to offset climatic changes and production may be attempted with the purpose to reduce water requirement by having a reasonably controlled canopy size and also to reduce the surface area exposed to transpiration.
- Develop package of practices to follow after the crop failure due to natural calamities like hailstorm/epidemics of pests and diseases.
- Development of integrated water and nutrient management (IWNM) practices coinciding with critical water and nutrient requirements of crop keeping in view the vagaries of climate.
- Working out "forecast models" for prediction of incidence of insect-pests and diseases and crop yield.
- Development of ideal rootstocks against biotic and abiotic stresses.
- Assessment of the carbon sequestration potential of citrus orchards.

CONCLUSION

Citrus is an evergreen subtropical crop, which has been acclimatized in tropical environments, too. In India, it is mostly grown in semi-arid regions of Andhra Pradesh and Maharashtra on vertisols. Under climate change and global warming, citrus production is likely to be benefited by elevated CO_2 and rise in temperature due to increased photosynthesis, growth, flowering, fruit retention, fruit weight and rapid colouration, provided other factors do not change. However, the crop is sensitive to indirect effects of climate change leading to increased frequency of aberrant weather and extreme events such as frosting, drought, flooding, unseasonal and erratic rainfall. Timely monsoon and evenly distributed rainfall have important bearings in rainfed citrus production influencing water stress, soil erosion, and disease and pest problems. Trend analysis of rainfall pattern in citrus-growing regions of Maharashtra indicates decrease in precipitation and thus water availability particularly in Vidharbha region. As a consequence of climate change particularly untimely rains at critical stages of the crop, aggravates the problem of insect-pests and diseases. Epidemics of citrus black fly, citrus psylla, citrus leaf miner, citrus rust mite are occurring with increasing frequency posing serious problem for citrus industry in India. Various adaptation measures have been evolved for management of these pests and diseases. There is an urgent need to further strengthen research to assess the impact of climate change and global warming on citrus in India and to develop appropriate location specific agro-techniques for adaptation.

References

Anonymous 2005. Annual Reports 2000-05, NRCC, Nagpur.

Anoymous. 2007. Vision 2025, NRCC, Nagpur.

Das, A.K. 2002. Pathogenic variability in *Xanthomonas axonopodis* pv. *Citri*, causal agent of citrus canker. *Journal of Mycology and Plant Pathology* **32:**247-48.

Das, A.K. 2008. *Phytophthora* diseases of citrus and their management. *Extension Bulletin* No. 10, NRC for Citrus, Nagpur.

Downton, W.J.S., Grant, W.J.R. and Loveys, B.R. 2007. Carbon dioxide enrichment increases yield of Valencia orange. *Australian Journal of Plant Physiology* **14:**493-501.

FAO. 2006. *Production Year Book*, 2006, Food and Agriculture Organisation.

Grierson, W., Cohen, E. and Kitagawa, H. 1986. *Fresh Citrus Fruit.* AVI Publishing, New York.

Huchche, A.D. 2008. Flowering, crop regulation and fruit drop in Citrus. *In: Five Decades of Citrus.* Singh, Shyam (Ed.), pp.188-203.

Idso, S. and Kimball, B. 2003. Effects of long-term atmospheric CO_2 enrichment on the growth and fruit production of sour orange trees. *Global Change Biology* **3:**89-96.

Joseph, C.V.C. Yoana, Vu, Newman, L., Hartwell Allen, Jr., Maria Gallo-Meagher and Mu Qing Zhang. 2002. Photosynthetic acclimation of young sweet orange trees to elevated growth CO_2 and temperature. *Journal of Plant Physiology* **159:**147-57.

Lurie, R., Porat and Gazit, Y. 2006. Control of Mediterranean fruit fly on citrus with catts. Technology. http://mbao.org/2006/06Proceedings/090LuriesS%20 Talk Summary.

Moss, G.I. 1969. Temperature effects on flower initiation in sweet orange (*Citrus sinensis*) *Australian Journal of Agricultural Research* **27:**399-407.

Panigrahy, P., Srivastava, A.K. and Huchche, A.D. 2006. *In-situ* soil and water conservation measures in Nagpur mandarin (*Citrus reticulata* Blanco) grown in central India. *Indian Journal of Soil Conservation* **34**:80-86.

Timmer, L.W. and Zitko, S.E. 2000. *Phytophthora* Brown Rot of Citrus: Temperature and moisture effects on infection, Sporangium production, and dispersal. *Plant Disease* **84:**157-63.

10

Impact of Climate Change on Productivity of Fruit Crops in Arid Regions

T. A. More and Rakesh Bhargava

THE earth is the only planet in our solar system which nurtures life. This is on account of fact that the earth has an atmosphere of proper depth and chemical composition. About 30% of incoming solar energy is reflected back to space and the remaining is used for maintaining an average temperature on earth. The chemical composition of earth atmosphere, which consists of 78% nitrogen, 20% oxygen and small amounts of CO_2 (0.036%) also helps to support life. The major role of atmosphere is maintaining life sustaining conditions on earth by not only absorbing energy reaching from the sun, but also by reflecting a part of it back to space in the form of infra-red waves. However, ever since the Industrial Revolution, man made activities have added significant quantities of GHGs to the atmosphere. The atmospheric concentration of GHGs of CO_2, methane and nitrous oxide has grown by about 31, 151 and 17%, respectively in last 150 years. Since these GHGs absorb the heat energy reflected from earth and retain them in the earth's atmosphere, the temperature of earth's atmosphere and in turn the surface temperature of the earth is gradually increasing. As a result of increase in GHGs the surface temperature has increased by 0.74 °C from 1906–2005. This phenomenon is called as climate change which can be defined as "statistically significant variation in either the mean state of climate or in its variability, persisting for an extended period".

The earth's climate system attempts to regulate the change by producing variation in cloud cover and change in wind direction and intensity. Thus, it introduces either positive feedback or negative feedback leading either to increase or decrease in the earth's surface temperature. It is estimated that by the end of this century, the earth's surface temperature will increase in the range of 2 to 4.5 °C along with changes in the rainfall pattern and increased frequency of extreme events of drought, flood and frosting, which will affect the agricultural scenario of the country. The fragile rainfed arid zone ecosystem supporting valuable horticultural fruit and vegetable crops of importance will become more vulnerable under climate change mainly due to rise in temperature and extreme events of frost and drought. The economy of the region is mainly based on animal husbandry, coupled with traditional rainfed farming which is not sufficient to provide stability for the present population of man and animals in the region.

CHARACTERISTIC FEATURES OF ARID ZONE IN INDIA

The arid region covers nearly 12% of land surface of India, spread over 39.54 million ha area, out of which 31.71 million ha is under hot arid region and 7.83 million ha under cold arid region. Indian arid zone is one of the largest subtropical deserts of the world of which 20% is arid and rest is semi-arid with varied habitats. The hot arid region occupies major part of north-western India including Rajasthan, Gujarat, Punjab and Haryana (73% of total arid region). The cold arid region mostly occurs in Jammu and Kashmir and Himachal Pradesh. The unusual features of arid ecosystem are extreme of temperature, low and erratic rainfall, low relative humidity, high potential evapo-transpiration and high wind velocity, etc. Soils are dry for most part of the year having aridic moisture regime and hyperthermic thermal regime. High temperature increases evapo-transpiration to an extent where water loss is considerable. The abundant sun shine and high light intensity increases transpiration through higher stomatal opening and high wind velocity remove moist air around the plants, disintegrates soil particle and thus causes soil erosion and sand dune formation. All these factors contribute to poor vegetation and low agricultural and horticultural productivity.

The natural vegetation have developed and/or modified their organs to perform certain vital physiological functions. Strong deep root system (ber, bael, aonla, wood apple, jamun etc.) to help forage moisture from deeper soil layers. Synchronization of flowering and fruit development with the season of moisture availability (ker, lasorda, aonla and pilu). Developed xerophytic characters like leaf shedding in summer (ber), scanty foliage (ker), spiny cladode (cactus pear), mucilaginous sap in plant part (ker, gonda, pilu and bael), sunken stomata and hairiness and waxy coating on the leaf surface (phalsa, ber, gonda and fig), thorny nature, and selective or reduced absorption of cation (Na^+) and anions (Cl^-, SO_4^{2-}) for survival under adverse aridic conditions. The scerophyllus shrubs, thorny forest, euphoria's, cacti and other xerophytic plants are marker for the arid ecosystem.

The arid zone ecosystem is very fragile and is prone to serious imbalances even with the slightest disturbance owing to mismanagement of resources or influx of external factors. Unscientific use of the scarcely available water in the face of recurrent droughts and the conditions of high potential evapotranspiration cause constraints to normal productivity. The poor fertility of desert soils and occurrence of high calcium carbonate further add to these constraints. The potential evaportranspiration in the arid districts of Rajasthan varies from over 2,000 mm at Jaisalmer to about 1,500 mm at Sikar with a moisture index value from -68.9 to -91.0. The arid districts in Gujarat record moisture index values from -70 to -82. In Haryana and Punjab, moisture index values vary from -59 to -73. In the peninsular arid region, the moisture index values of -68.6 to -70.0 are common. These conditions are caused as a result of the extremely high temperatures, hot winds and high solar radiation during summer and very low rainfall. Obviously, atmospheric vapour pressure deficit reaches as high as 30 mb during the summer. The irrigation water resources in these regions are also limited. Therefore, the success largely depends on efficient management to ameliorate the extreme agroclimate and on optimizing use of water.

Despite the fact that arid zone agroclimate is harsh, it is conducive for high quality production of horticultural crops. The peculiar agroclimatic conditions impart unique quality in fruits such as ber, kinnow, bael, date palm, pomegranate, aonla, papaya, and vegetables such as chilli, fenugreek, cucurbits and spices crops like cumin. Arid region is the only area where commercial date palm production is possible. Such conditions offer great opportunity for production for export of these crops. The indigenous products from *Capparis* and *Cucumis* can be exploited for preparation of value-added products for internal as well as export consumption. The high solar radiation resource provides opportunity for its harvesting by optimation of cropping system models incorporating multilayer receptor crops. This also offers opportunity for insurance against high risk conditions of the arid zone. The availability of considerable amount of surplus family labour in the arid region farming communities is also a great resource to be put to use for increasing production of high-value horticultural crops.

CLIMATE CHANGE IN ARID REGIONS OF INDIA

Since, last few years the arid region is also facing the variation in climatic parameters. The data compiled for last 20 years of Bikaner region show that there is a shift in maximum and minimum temperaturer in this region. Perusal of recent weather data indicates that yearly maximum temperatures remained nearly constant for last 20 years, however, the spectrum of minimum temperature shows that prior to year 2000 the minimum temperature never touched 0 °C, but since then it has become a recurrent feature that the temperature goes below 0 °C. Similarly, weekly mean maximum temperature showed a range from 42.3–46.4 °C. The mean maximum temperature of 46.4 °C reached only in 1998, whereas during other years it remained around 45 °C. Similarly, the mean weekly minimum temperature revealed a drastic change in last 10 years. The mean weekly minimum temperature remained between 4.7–5.2 °C up to 1999, however, since then the area is experiencing dip in mean weekly minimum temperature. It dropped to around 3 °C in 2000 and 2001, around 2 °C in 2003, 2004 and 2007 with the lowest 1.8 °C in 2006.

The lowest temperature in Bikaner region is recorded mostly in January. The data of past three years shows that during 2006 the minimum temperature dropped to sub-zero level and persisted for continuously for 3 days. The same pattern was repeated in 2008 also and this year there were two dips firstly in the first week of January and secondly by 20th January when the temperature dropped to -0.5 °C and remained so for 2-3 days.

The total rainfall pattern in Bikaner remains highly unpredictable since a wide variation in total rainfall was recorded in different years. During last 20 years, maximum annual rainfall (472.1 mm) was recorded in 1997 and minimum (92.8 mm) was recorded in 2002. In other years, the annual rainfall remained between 130–315 mm. Thus, the change in relative humidity (RH), number of rainy days and total rainfall does not show any specific pattern during last 20 years. In most of the years the rainfall remains less than the average for the region leading to recurrent drought, this is the characteristic feature of arid ecosystem.

Thus, can be argued that due to overall change in climatic parameters at global levels, the arid region has experienced: (i) a shift in minimum temperature which goes below 0 °C and persist for 2-3 days leading to frost injury in plants, and (ii) irregular rainfall pattern leading to recurrent droughts have become a common feature in this region.

IMPACT ANALYSIS ON ARID FRUIT CROPS

Elevated CO_2 and Temperature

Productivity

The impact of elevated CO_2 levels has received very less attention. Only limited studies are available to demonstrate the effect of elevated carbon dioxide on growth and fruit quality. It has been postulated that the productivity of arid horticultural crops will increase under elevated CO_2 level provided other parameters such as irrigation, temperature, etc., remains non-limiting.

A study conducted on growth and development of aonla seedling under elevated CO_2 level demonstrated that the relative growth rate (RGR), net assimilation rate (NAR), CO_2 assimilation rate, R:S ratio and water-use efficiency increased under elevated CO_2. However, positive impact of elevated CO_2 was down regulated beyond 30 day exposure. Specific leaf area (SLA), transpiration rate, stomatal conductance declined due to exposure to elevated CO_2 (Khurana and Singh, 2005).

Quality

The studies conducted on storage of pomegranate fruits have demonstrated that if the fruits are stored under modified environment having 10 or 20 kpa CO_2 at 10 °C, the aril becomes light pink in colour. This is due to that the synthesis of anthocyanin decreased under high CO_2 level (Holctroft *et al.*, 1998). Similarly, an experiment conducted using date palm fruits of cv. Barahee demonstrated that exposure of fruits to 20% CO_2 significantly increased shelf-life and maintained TSS. Elevated CO_2 further retarded the degradation of Caffeoylshikimic acid, an important tannin in date fruits (Al-Redhaiman, 2005).

Pests of Target Crop

The information on effect of elevated CO_2 and rise on temperature on arid zone horticultural crops is not readily available.

ABERRANT WEATHER AND EXTREME EVENTS

On account of aberrant change in climatic conditions, horticultural scenario of the region has experienced a drastic change in terms of survival of crops and productivity. Some of the common impact of climate change are summarized crop-wise as under.

Aonla: Aonla (*Emblica officinalis*) fruit is valued high among indigenous ayurvedic medicines for balanced diet and sound health. The fruit contains a chemical substance known as polyphenols which retards the oxidation of ascorbic acid. Therefore, aonla is a rich source of vitamin C (630 mg/100g fruit) next to barbados cherry (*Malphigia glabra* L.) in fresh as well as dried form. The aonla was introduced in arid region for commercial cultivation and it showed its potential in normal years without frost. Since, the crop is highly susceptible to frost, its commercialization in frost zones of arid region is now a debatable issue. The complete plant gets badly damaged due to frost and the recovery is very poor in graft scion. It has been experienced that on the event of frost, aonla crop is severely affected and in young budded plants it has been observed that frost effect to the scion portion and dried the young grafts. It has also been experienced that aonla matures before occurrence of frost and maximum produce is harvested. However, fruits of late maturing varieties gets affected by frost and low temperature (< 2-3 °C). The fruits become whitish in colour, water starts oozing out of them and subsequently they dry and turn black. The plant growth developed after frost injury is poor and particularly in graft scions which leads to no flower/fruit production in next crop season. Under extreme low temperature the whole plant is killed.

Ber: Ber (*Ziziphus mauritiana*) is common and native plant of arid region. The plant is drought hardy and hence survives and gives economic yield in the region. The fruits are nutritious and hence it is also called as the desert apple. Despite the fact that ber is native of arid region, recently the plant is showing some morphological and physiological changes in response to low temperature, or irregular rainfall being experienced by the region. The ber plants can survive a minimum temperature of 4 °C and can tolerate maximum temperature of 42 °C. However, various physiological processes gets affected at different temperature regimes. For instance, optimum temperature for seed germination is 30 °C and germination gets inhibited if temperature goes below 25 °C or above 35 °C. Similarly, fruit setting is drastically gets affected if temperature exceeds 35 °C.

The response to low temperature of crop is highly variable and cultivar specific. During 2005, 2006 and 2007, it has been observed that out of 311 ber genotypes planted in National Repository of Ber at CIAH, Bikaner research farm, cultivars Tikkadi, Syriya and Sanaur-5 showed tolerance to frost and do not get affected with frost, whereas Umran, Mundia and Aliganj were highly susceptible in which about 50–60% plants get damaged. In susceptible cultivars, leaves, fruits and branches gets affected in outer periphery of canopy, which leads to loss of about 40% pruned wood weight in crop. The cultivar Sanaur-1, Jogia and Kathaphal are moderately (25–30%) affected by frost. Since, frost normally coincides with fruit maturity period; fruits get badly affected and cause a heavy loss in majority of cultivars. On frost injury, fruits get shriveled, become brown that later turns into black colour and finally dry and drop. As a result of this, the yield reduction is 30% in Rashmi, 20% in Seb, 52% in Umran, and 60% in Illiachi. The recovery of plants after frost injury is normally good in this crop, but repeated effect of frost leads to disturbed carbohydrate metabolism which ultimately effect vegetative vigour and reproductive biology of plant.

The flowering in this crop is not affected by frost but irregular rainfall has bearing on this process. The flower initiation which normally takes place during second week of August to last week of September gets delayed and extends up to October and flower drop enhanced up to 35–40%.

Pomegranate: Pomegranate (*Punica granatum*) is one of the favourite table fruits in tropical and subtropical regions. It is considered as a crop of famine area because of its ability to withstand the soil and climatic conditions. The juice is considered to be useful for patients suffering from leprosy. Rind of the fruit is commonly used in dysentery and diarrhoea. The plant can grow best in a temperature range of 10–35 °C. An analysis of climate change on pomegranate in arid region has demonstrated that the plant is highly affected by the low temperature. In this crop, the leaves and young shoots are badly affected by frost. Thus, plant growth and development are highly affected leading to no flower production in next bahar. On the event of frost injury in the crop, the complete foliage of the plant is dried within 2-3 days and the plant is defoliated. The young shoots also dries killing all vital buds on twig. Therefore, after leaf shedding, removal of affected twigs is essential for renewed growth of plant. The recovery of plant after injury with frost is fast and the plant can bear flowers and fruits only after 6 months.

Under normal conditions, the plant has potential to flower throughout the year. Keeping in view the climatic conditions of the region, the commercial production can be harvested in one of the three seasons, *viz. ambe*, *mrig* and *hasta bahar*. In arid conditions, it was generally practisesed to take the crop in *mirg bahars*. This was on account of fact that the flowering and fruit development coincides with the rainy season and fruit mature during winter when the temperature is congenial for development of colour in arils. However, with the introduction of frost in this region, in this *bahar* the fruits gets hardened, dull skin colour, finally rot and become unmarketable at maturity. If other bahars, i.e., *hasta* and *ambe* are adopted for cultivation of this crop in arid region, the colour do not develop in arils due to high temperatures (>40 °C). In addition to this, fruits developed under *hasta bahar* also get affected by the frost leading to heavy economic loss. Under these conditions, it is necessary to standardize the bahar period for obtaining economic yield in arid region.

The effect on fruits is characterized by the initially hardening of fruit due to freezing and subsequently it becomes pulpy on account of thawing. The fruit finally become black due to rapid infection of pathogen.

Date palm: Date Palm (*Phoenix dactylifera)* is considered to be the oldest cultivated tree fruit. It is believed to be indigenous to countries around Persian Gulf. There are three major species of date palm grown in India namely, *Phoenix dactylifera*, *P. sylvestris* L. Roxb and *P. canariensis* Chabaud. *Phoenix dactylifera* is very common and grown commercially because of its superior quality edible fruits. The plant has immense economic value as each and every part of the plant is used for value addition. The CIAH, Bikaner, is having the National Repository of date palm in which a total of 55 accessions are available. Although the optimum temperature for the growth of this crop species is 32 °C, the zero vegetation point for the crop is 7 °C and above 40 °C the growth of the plant stops.

The flowering of date palm only takes place after a brief exposure to low temperature. The effect of frost on the date palm cultivars was studied and it was found that the crop is only slightly affected by frost. The plant does not show any influence on its morophological parameters and growth remains normal. However, it was observed that the spathe emergence, flowering and pollination is delayed if plants experience low temperature for a longer period. The spathe emergence in date palm normally takes place during last week of January to mid of February but if the plant has experienced low temperature then it gets delayed and spathe emergence takes place between first week of February and extends up to first week of March. In response to this, the fruit maturity is also delayed which extends up to mid-July. This may be on account of fact that decrease in temperature leads to late accumulation of required degree days for the physiological activity. Minor increase in flower drop has been recorded and yield reduction to the tune of 10% has been noticed. Prolong low temperature also hinders the spathe emergence and flowering in male palm. Ultimately the pollination of female palms does not take place at proper time.

Bael: Bael (*Aegle marmelos*) is an important indigenous fruit having mythological significance. The importance of bael lies in its curative properties, which make the tree one of the most useful medicinal plants of India. The roots as well as fruits of this species is used for preparation of Ayurvedic medicine. The *Aegle* consists of 3 species of which only one is cultivated. The tree is deciduous, hardy and have an ability to withstand in arid conditions, saline soils (6–9 dsm^{-1}) and sodic soils (20-30 ESP). The main bael-growing states are Uttar Pradesh, Bihar, West Bengal, Madhya Pradesh, Jharkhand and Orissa. The cultivation of bael in the arid region was attempted and at present a total of 21 accessions of bael are available in the germplasm at CIAH, Bikaner. It has been reported that bael plants are relatively frost tolerant and are not affected even at low temperature of -7 °C. The effect of frost on the bael was studied and it was observed that the 50% of its young plantation burn in response to frost but the plant recovers speedily.

Kinnow: Kinnow is very well adapted in arid region with irrigation facilities. In view of this the kinnow is growing well under irrigated conditions of arid region. The production of crop is very good because of high temperature and low humidity. The crop grows well in the temperature range of 10–40 °C. The crop matures during December–January. The frost also affects the young growth of plant but its fruits are not affected. The young twigs are affected by frost which dries and are required to be pruned for growth and development of the plant. It has been observed that after frost, the new growth in plant gets accelerated as soon as the temperature rises above 20 °C. Since, economic yield of the crop is not affected by the change in the climate particularly due to frost, the crop has shown its potential to be an ideal for irrigated parts of hot arid region.

Karonda: Karonda (*Carissa carandus*) is an evergreen, densely branched, spiny shrub, proves to be an excellent bio-fencing material for raising hedges. It is widely grown in India, South Africa and Malaya. In India, it is found growing wild in Bihar, West Bengal, south eastern part of Chhattisgarh, parts of Orissa, Maharashtra, Aravali hills and southern part of India. In Bastar, a tribal tract of Chhattisgarh, white and red fleshed karonda fruits weighing about 30-40 g have been located. There are about 20 species of which 5 are of Indian origin including *C. carandus*. In arid region, karonda has been introduced as part of crop diversification. The plant is highly susceptible to frost and its young leaves and shoots gets burned (40–50%) with freezing temperature. The plants recovers fast and get early flowering and fruiting in the ensuing season.

Lasora: Lasora (*Cordia myxa*) is very well adapted crop of arid region. On the event of frost, lasora is the first crop in the region which gets affected and initially the leaves curved, subsequently dries. Higher frost period affects young and soft twigs which finally required to be pruned.

Thus, from the foregoing account it is imperative that frost and irregular rainfall are the major climatic parameters which have changed in last few years. The arid fruit species are getting affected because of this and hence there is a need to develop appropriate strategies for tiding over the present situation. This calls for research preparedness so that technologies for mitigation of adverse effect of frost can be overcome. Recovery of the plant is very fast as soon as the temperature rises. Since the fruiting season of crop is March–April, crop is not affected much in terms of its economic potential.

Adaptation Measures for Climate Change and Global Warming

- Identification of stress tolerance genes in the indigenous plants of arid region.
- Develop heat, frost and drought tolerant cultivars of arid horticultural crops.
- Development of cultivars with high carboxylation and water use efficiency.

- Development of agro-techniques for cultivation of arid horticultural crops under changing climate scenario.
- Introduction of new crops to fit into the cropping models and suitable under changed climatic conditions.
- Development of pest and disease forecasting models.
- Development of fertigation techniques to economize use of water and maintain micro-environment.
- Human resource development to educate the farmers regarding measures to be adopted.

Mitigation Measures to Reduce Emission of CO_2/GHG and Carbon Sequestration Potential of Crops

- Carbon sequestration potential of different cropping models needs to be evaluated for adoption under arid region.
- Development of cropping models which can optimally utilize CO_2 and reduce its emission.
- Identification of microbes which can help in reducing GHGs.
- Use of slow release/soluble fertilizers in agriculture.
- Resource conservation techniques.
- Organic farming.

FUTURE STRATEGIES FOR RESEARCH

- ***Identification and evaluation of genotypes from other regions for frost tolerance:*** Due to change in climatic conditions of the arid region, frost has become a regular feature in this region. Even one day dip in temperature below zero degree centigrade can spoil the whole crop of aonla, lasora, karonda, pomegranate *etc*. In a situation such as this, the need of the hour is to identify the genotypes of the said crops having tolerance to frost from the other regions where the frost is a regular feature. This material can either be used directly as new genotype or can be used as breeding material to impart tolerance in existing commercial cultivars of respective crop.
- ***Breeding cultivars of arid horticultural crops tolerant to abiotic stresses of heat, drought, salinity and frost.***
- ***Development of agro-techniques for overcoming adverse effects of frost:*** Since frost is a regular feature in arid region, there is an urgent need to develop agro-techniques for mitigation of effect of frost. This involves development of appropriate agro-techniques for mitigation of the effect of frost. This includes cultural practices such as applying light irrigation, spray of sulphuric acid, smoking during night hours, use of electric heaters, application of mulch etc. for protection of crop during frost.
- ***Development of multi-strata farming system models for sustainable crop production under climate change situation:*** In arid and semi-arid regions, mono culture is a risky proposition. Hence, it is recommended to adopt appropriate multi-strata farming system for sustainable crop production. It has been demonstrated that in the event of the under storey crop remains protected by the over storey component and the economic loss is

drastically reduced as compared to that grown as sole crop. Thus, there is a need to develop multi-strata cropping models using multi-tier system taking into consideration fruit crops, vegetables, medicinal crops for sustainable crop production.

- ***Crop diversification by introduction of new crop species:*** New crop species from regions where frost is a regular feature needs to be identified and introduced in the arid region which are either tolerant to frost or escape frost by early fruiting and maturity.
- ***Studies on the insect and pest population dynamics under climate change.***
- ***Appropriate strategies are to be developed to mitigate the effect of irregular rainfall:*** The rainfall pattern in the arid region is low and irregular. At times, there are normal rains during sowing time and subsequently the next shower takes place after a long gap. In a situation such as this, there is a need to develop appropriate strategy to save the crop from drought period. The studies will be undertaken on seed priming, application of chemicals such as BAP, thiourea, salicylic acid *etc.* to save the crop, development of water storage tanks for life-saving irrigation, application of microirrigation to minimize the water loss, use of drought tolerant cultivars, identification of rootstock for different fruit crops. Studies on plant spacing and moisture conservation techniques need to be standardized.

CONCLUSION

Human activities during last century have changed the composition of the atmosphere leading to aberrant changes in climate. Over past one century, on account of Industrial Revolution, a quantum of GHGs have been added to the earth's atmosphere which increased the earths surface temperature. Although, at all India level no significant trend has been observed in rainfall pattern, but at localized zones the impact has become visible. The Indian arid region has also experienced a shift in maximum and minimum temperature in last two decades. It has been observed that the minimum temperature dip to sub zero level. This has affected the production and productivity of arid horticultural crops, as most of these crops are highly sensitive to frost. Arid zone horticultural crops are also very sensitive to irregular rainfall pattern affecting moisture availability at critical stages. There is an urgent need for research to develop adaptation measures, introduce and breed new cultivars to sustain the productivity of the arid horticultural crops.

References

Al-Redhaiman, K.N. 2005. Modified atmosphere extends storage period and maintains quality of Barhi date palm. *Acta Hort.* (ISHS) **682:**978-86.

Holcroft, D.M., Gil, M.I. and Kadar, A.A. 1998. Effect of carbon dioxide on anthocynanins, phenylalanine ammonia lyase and glucosyltranferase in the arils of stored pomegranate. *J. Amer. Soc. Hort. Sci.* **123:**136-40.

Khurana, E. and Singh, J.S. 2005. Response of five dry tropical tree seedlings to elevated CO_2: Impact of seed size and successional status. *New Forest* **27:**139-57.

11

Climate Change and Seed Spices

O.P. Aishwath, S.S. Meena, K. Kant, Y.K. Sharma and M.M. Anwer

SEED spices are annual crops, giving dried fruits or seeds used as spices namely coriander, cumin, fennel, fenugreek, *ajowain*, dill, celery, aniseed and nigella (*kalonji*). These are characterised by pungency, strong odour, sweet or bitter taste used in whole or ground form for imparting flavour, aroma and pungency for seasoning of food. The value-added products of seed spices are — spices oil, oleoresins, ground spices, curry powder, consumer packed spices and organic spices. The major seed spices-producing states are Rajasthan, Gujarat, Chattisgarh, Madhya Pradesh, Andhra Pradesh, Tamil Nadu and Uttar Pradesh. The prevailing worldwide demand for seed spices is estimated to be 1,50,000 tonnes. Of which India contributes 83,550 tonnes annually, accounting for 55.7% of the total world trade, valued at Rs 275 crore. The seed spices account for about 36 and 17% of the total area and production of spices in country with significant export value.

Climate change and global warming, caused by greenhouse gasses and industrial emissions is estimated to increase the annual mean earth surface temperature ranging from 3 to 5 °C by the end of century. The warming will be pronounced in the northern parts followed by west coast and west central India associated with increasing frequency of aberrant weather and extreme events of flooding, drought and frost. Therefore, it is imperative to assess the impact of climate change on seed spices for timely adaptation measures and evolve strategies for sustained production of high export value seed spices in India.

IMPACT ANALYSIS ON SEED SPICES

Elevated CO_2 and Temperature

Information on the effect of elevated CO_2 and temperature on seed spices are severely lacking.

Productivity

Indirect evidences suggest beneficial effect of elevated CO_2 and rise in temperature on seed spices, which needs to be verified. Most of the seed spices belong to carrot family and may show the similar response. The carrot yield increased with the rise in temperature and CO_2 (Wheeler *et al.*, 1996).

Seed spices are winter season crops that require cool climate for their growth and development, but contrasting warm weather conditions with bright sunny days for their maturity. However, very low temperature or frosting leads to the severe crop damage. In general, marginal and sub-marginal lands are utilized for the production of these crops. The area and production during last 10 years also does not show any consistent trend in production of seed spices. The area and production of coriander, cumin, fennel and *ajowain* tended to increase, whereas that of fenugreek fluctuated widely and on an average remained static. Indirectly, it shows that coriander, cumin, and fennel are perhaps more adoptive to environmental conditions than fenugreek and *ajowain*. Moreover, the demand of *Ajowain* is not stable in the market, affecting its area and production.

Quality

Information on the effect of elevated CO_2 and temperature on quality of seed spices is severely lacking. The occurrence of rainfall during March and April deteriorates the quality of seed spices (physical, chemical and biological properties and reduce oil content, etc.).

Pests of Target Crop

Midge (*Systole albipennis*) is the most important pest of coriander. Thrips attack coriander and cumin, whereas aphids are a problem in coriander, fennel and fenugreek. High temperatures reduce midge and thrips population indicating reduced infestation under climate change and global warming. The cloudy weather during January and February leads to heavy infestation of aphids in seed spice crops and outbreak of powdery mildew disease in cumin resulting in heavy losses. Winter shower at the time of flowering invites a large number of insect pests and diseases and high crop damage occurs. Cumin is highly susceptible for blight aggravated by the winter showers. Aphid outbreaks in seed spices are associated with temperature rise from 5 °C to 10 °C. Hence, there is a possibility of heavy infestation of aphids in coriander, fennel and fenugreek with global warming in future.

OTHER INDIRECT EFFECTS OF CLIMATE CHANGE

Salinity and Alkalinity Stress: Seed spices are grown in arid and semi-arid regions of India. In these areas, soil salinity is a widespread and serious problem. This problem will further aggravate with climate change due to high evapotranspiration that brings out the salt on surface. Most of the seed spices have the adaptability for salinity and alkalinity but further screening for their tolerance level to combat against the rising problem of salt stress is required. Some examples are given here for the upper tolerance limit of salt.

Some of the seed spices crops like *dill* seed (*Anethem sowa*) and celery (*Apium graviolens*) could be plasticized with available poor quality of under groundwater (Dagar, 2007). The coriander variety CO-2 was reported as salinity and alkalinity tolerant and can be cultivated in arid and semi-arid areas for the production of coriander essential oil (Aishwath *et al.*, 2008). *Dill* can be grown up to EC 3.3 ds/m (Sharma *et al.*, 2001). In another study, it was found that *dill* (*Anethem graveolens*) can tolerate pH up to 8.6 (Farooqi and Sreeramu, 2001). Fennel and coriander grown on partially reclaimed sodic soil having Exchangeable Sodium Per cent (ESP)

20 with various N and P levels yielded well and large accumulation of Na in fennel stover indicating its tolerance to sodicity. Reduction in soil pH and ESP and increase in exchangeable calcium indicates its capability for reclaiming sodic soils.

A decline in biomass for fenugreek and coriander was recorded at ECe 7.3, 5.2 and 3.3 ds/m, respectively (Sharma *et. al.,* 2001). The increase in suppression of seed germination of cumin with the increase in salinity was observed in all the nine cultivars tested. The genotype UC 208 showed maximum salt tolerance, followed by UC 209 and UC 218 (Dhayal *et al.,* 2001). Both salinity levels, root length and harvest index exhibited positive direct effects on yield, while branches/plant and protein content had negative effects on yield (Mathur *et al.,* 1999).

Aberrant weather and Extreme Events

- During 2007-08, heavy yield loss of coriander, cumin, fennel and *kalijeeri* occurred by the frost at the research farms and farmers' fields in Rajasthan. Sudden rise in temperature during January and February advanced the early maturity of crops. The early maturity leads to improper seed setting and seed filling. Ultimately growth period reduced and about 50% yield reduction took place. It shows that these crops are highly sensitive to aberrant weather condition.
- The *kalijeeri* crop was totally damaged with frost during 2007-08 in Rajasthan (Fig. 11.1).

Fig. 11.1. Crop of *kalijeeri* (*Centeratherum antihelminticum*) failure by frost damage during 2007-08 in Rajasthan

Adaptation Measures for Climate Change and Global Warming

- Shelter and windbreaks to protect from possible frosting as most of the seed spices are winter season crops and are sensitive to frosting.
- Use of drip and sprinkler irrigation to save water.
- Biological control of pests.
- Some drought tolerant varieties are CO-1, CO-2, Sadhana CS-4, Swathi (CS-6), RCr-2 and RCr-436 for coriander, NRCSS-AA-1 for Ajowan and NRCSS-AD-2 for *Dill* (Aishwath *et al.*, 2008).
- Drainage and soil conservation techniques with emphasis on improving soil organic matter, because seed spices in marginal light textured soils are sensitive to both flooding and drought.
- Though nutrient requirement of seed spices is small, they are grown in marginal soils with poor fertility.
- Inclusion of short duration seed spices in cropping systems for increased efficiency and profitability.

Mitigation Mesures to Reduce Emission of CO_2/GHG and Carbon Sequestration Potential of Crops

Seed spices being annual crops have very little carbon sequestration potential. Other mitigation measures are as follows:

- Resource conservation techniques
- Organic farming.

FUTURE STRATEGIES FOR RESEARCH

- Seed spices are in general short duration crops, therefore, short duration varieties should be developed to fit into the other cropping systems.
- Screening genotypes of seed spices with tolerance to heat, drought, salinity, alkalinity, nutrients, frost and pest stress.
- Developing precision and organic farming techniques.
- Value-added product development like fenugreek as biodegradable flocculating agent for industrial effluent, seed spices straw for nylon production *etc.*
- Development and validation of seed spices crop simulation model.

CONCLUSION

Seed spices are annual crops, giving dried fruit or seeds. These crops grown on marginal lands with light textured soils in India have high export value. These are relatively hardy crops with small nutrient and water requirement, but are sensitive to flooding, drought and frost, but relatively tolerant to salinity and alkalinity. Seed spices crops not only have export earning potential, but also could play a vital role in the environment protection and climate change. These crops appear to be better equipped to adopt in changing environment with potential for utilization in problematic shallow soils with proper agronomic management. Climate change and global warming may impact the productivity of these crops as they are mostly grown in arid and semi-arid regions vulnerable to warming. However, information is lacking on the response of these crops to elevated CO_2 and rise in temperature, predicted under future climate change scenarios. The information needs to be generated urgently to utilize the potential of these crops under climate change and global warming to sustain the production of these high-value export-earning crops through adaptation and relocation in other areas and cropping systems.

References

Aishwath, O.P., Vashishtha, B.B. and Malhotra, S.K. 2008. Scope of production of aromatic plants in arid and semi-arid regions. *In: Proceedings of National Workshop on Spices and Aromatic Plants*, held on 6-7 February 2008 at ARS, RAU, Mandor, Jodhpur, Rajasthan, pp. 10.

Dagar, J.C. 2007. Alternative land uses in areas with water scarcity. *In: On-Farm Land and Water Management (A compilation of National Training Course)*, pp. 154-65.

Dhayal, L.S., Bhargava, S.C. and Jakhar, M.L. 2001. Salinity tolerance of cumin (*Cuminum cyminum* L.) genotypes during germination. *Journal of Spices and Aromatic Crops* **10:**145-46.

Mathur, V.L., Ladu Lal and Lal, L. 1999. Association analysis under saline conditions in Fenugreek (*Trigonella foenum graecum*). *Indian Journal of Agricultural Research* **33:**227-32.

Sharma, P.C., Mishra, B., Singh, R.K. and Singh, Y.P. 2001. Variability in the response of spinach, fenugreek and coriander to alkalinity and salinity stresses. *Indian Journal of Plant Physiology* **6:**329-33.

Wheeler, T.R., Ellis, R.H., Hadley, P., Morison, JIL, Batts, G.R., Daymond, A.J. and Williams, R.J. 1996. Assessing the effects of climate change on field crop production: Implications of 'Global environmental change' for crops in Europe. *Aspects of Applied Biology* **45:**49-54.

12

Impact of Climate Change on Potato in India

J. P. Singh, S. S. Lal, P. M. Govindakrishnan, V. K. Dua and S. K. Pandey

CLIMATE change is now an acknowledged fact and reality. The evidence gathered world over using state-of-the-art technology by various national and international agencies is irrefutable. Human activities like rapid industrialization, intensive agriculture, indiscriminate use of fertilizers, deforestation and increasing use of fossil fuels during past 150 years are the major contributing factors for climate change. The continued effect of these activities resulted in increasing emission of CO_2 and other greenhouse gases (GHG) leading to global warming as a '*greenhouse effect*' due to entrapment of back radiation from earth by these gases. The increase in temperature due to global warming is 0.76 °C since 1850. The rate of warming in last 50 years is double than that for the last century. As many as 11 of the past 12 years were warmest since 1850, when records began. The likely increase in temperature is 1.8–4 °C by next century. The CO_2 concentration is projected to double from the current level of 360 ppm in the atmosphere. Global warming is occurring along with shifting pattern of rainfall and increasing incident of extreme weather events like floods, droughts and frost.

The growth of potato in India has been phenomenal since 1950 with increase in area production and productivity by 6, 15 and 3 times, respectively. Globally India stands at fourth and third position now with respect to acreage and production, respectively. The crop is mainly confined to Indo-Gangetic plains in mild and cool winters in India. The autumn/winter planted crop in northern plains of India comprising Uttar Pradesh, West Bengal, Bihar, Punjab and Haryana contributes 84% of total potato production in India. Here, the crop is grown totally under irrigated conditions. It is also grown in small scattered areas as rainfed crop in hills during summers and as rainy (*kharif*) and winter seasons crop in plateau region. However, climate change and global warming will have a profound effect on potato growth in India impacting every aspect of not only production and profitability, but seed multiplication, storage, marketing and processing of this perishable vegetatively propagated crop. Under the impact of future scenarios of climate change, the growth projections of potato in India might be arrested or even reversed, unless effective adaptation measures are evolved for timely application.

IMPACT ANALYSIS ON POTATO

Increase in temperature and atmospheric CO_2 are interlinked occurring simultaneously under future climate change and global warming scenarios. Effect of their interaction on potato would be more relevant and of greater economic significance compared to their usually counteracting direct effects on crop growth, yield and quality. Potato is mostly grown in north India during winters usually receiving few scattered rains. Under future scenarios, the global warming is projected to be more pronounced over land areas with maximum temperature increase over northern India. The winter and post-monsoon seasons are likely to be more affected by warming. Therefore, potato in addition to direct effects on growth and yield may be subjected to indirect effects of warming. These are increasing drought due to reduction in precipitation accentuating salinity and unpredictable extreme events of erratic unseasonal rains, flooding and frosting *etc.*

Elevated CO_2 and Temperature

Productivity

All India estimates of potato production were made by INFOCROP-POTATO simulation model (Singh, 2005 and 2008) without adaptations and assuming that area under the crop remains constant at current levels (1.2 m ha) in future climate scenarios (Table 12.1). Results showed that potato production will increase by 11.12% at elevated CO_2 of 550 ppm and 1 °C rise in temperature. However, future climate scenarios for India indicates that at elevated CO_2 of 550 ppm temperature rise is likely to be 3 °C (IPCC, 2007), with decline in production by 13.72% during 2050. The 1 °C rise in temperature is likely to be associated with only 400 ppm of CO_2 to be assumed in 2020 (IPCC, 2007), with a decline in potato production by 3.16% (Table 12.1).

Table 12.1. Change (%) in potato production in India from current levels as affected by elevated CO_2 and rise in temperature without adaptations

Atmospheric CO_2	*(ppm)*	*Rise in temperature (°C)*				
	nil (current)	*1(2020)*	*2*	*3(2050)*	*4*	*5(2090)*
369 (current)	0.0	–6.27	–17.09	–28.10	–42.55	–60.55
400 (2020)	3.40	–3.16	–14.57	–25.54	–58.63	–58.63
550 (2050)	18.65	11.12	–1.25	–13.72	–30.25	–49.94

(Values in parentheses are likely years for associated CO_2 levels and temperature rise)
Source: Singh, 2009

***Direct effect of elevated CO_2*:** The effect of elevated CO_2 concentration in controlled experiments conducted in OTC (Open Top Chambers), FACE (Free Air Carbon Enrichment) and growth chambers overwhelmingly suggests positive effect on growth and yield with only few negative influences.

The CO_2 concentration and assimilation are positively correlated. Doubling the CO_2 concentration from ambient level of 360–720 ppm increased total biomass by 27–66% (Collins, 1976; Heagle *et al.*, 2003; Olivo *et al.*, 2002; Donnelly *et al.*, 2001; Miglietta *et al.*, 1998; Van de Geijn & Dijkstra, 1995). The tuber yield increased from 32–85% (Collins, 1976; Wheeler *et al.*, 1991; Heagle *et al.*, 2003; Craigon *et al.*, 2002; Olivo *et al.*, 2002; Donnelly *et al.*, 2001; Miglietta *et al.*, 1998; Finnan *et al.*, 2002). The increase in tuber yield is estimated to be approximately 10% for every 100 ppm increase in CO_2 concentration (Miglietta *et al.*, 1998). These positive effects are attributed to increased photosynthesis from 10 to 40% (Collins, 1976; Katny *et al.*, 2005; Olivo *et al.*, 2002; Vandermeiren *et al.*, 2002; Schapendonk *et al.*, 2000). The increase in photosynthesis was most marked in young leaves (Katny *et al.*, 2005; Vandermeiren *et al.*, 2002). This is attributed to phenomenon of photosynthetic acclimation later in the growing season particularly in old leaves (Vandermeiren *et al.*, 2002; Schapendonk *et al.*, 2000; Lawson *et al.*, 2001). Varietal differences in response to elevated CO_2 concentration exists (Olivo *et al.*, 2002).

Number of tubers remained unaffected under elevated CO_2, but mean tuber weight increased mainly through increase in number of cells in tubers without influencing the cell volume (Collins, 1976; Chen and Setter, 2003; Donnelly *et al.*, 2001). However, an increase in tuber number has been also observed (Miglietta *et al.*, 1998; Craigon *et al.*, 2002).

Elevated CO_2 concentration reduces evapotranspiration (ET) resulting in water saving to the extent of 12 to 14 % (Magliulo *et al.*, 2003; Olivo *et al.*, 2002).

Elevated CO_2 concentration advances the tuber initiation and flowering (Miglietta *et al.*, 1998) but hastens senescence of leaves (Miglietta *et al.*, 1998; Vaccari *et al.*, 2001)

The few negative effects of elevated CO_2 concentration include reduction in chlorophyll content in leaves particularly during later growing season after tuber initiation (Bindi *et al.*, 2002; Lawson *et al.*, 2001).

***Direct effect of temperature*:** Growth and development are affected at high temperatures encountered in the tropics. No potato crop growth is possible below 2 °C and above 30 °C (Van Keulen and Stol, 1995). The minimum (0–7 °C), optimum (16–25 °C) and maximum (40 °C) temperatures for net photosynthesis are reported (Kooman and Haverkort, 1995). Potato requires cool night temperature to induce tuberization (Burt, 1964; Ku *et al.*, 1977; Cutter, 1992). Although photosynthesis in potato is suppressed by high temperature (Ku *et al.*, 1977), it is not as sensitive to temperature as tuberization and partitioning of photosynthates to tuber (Reynolds *et al.*, 1990; Midmore and Prange, 1992). The radiation-use efficiency (RUE) is suppressed under high temperatures (Allen and Scott, 1992). High temperature reduces tuber number and size (Ewing, 1997).

High temperature brings about marked morphological changes like etiolated growth with smaller size of compound leaves and leaflets reducing the LAI (Ewing, 1997; Fleisher *et al.*, 2006) in addition to reduction in tuber number and size (Peet and Wolfe, 2000; Khan *et al.*, 2003). However, long day conditions and high temperature prevailing in spring season in Punjab favoured growth of foliage at the cost of tubers and improved processing quality of tubers (Marwaha and Sandhu, 2002).

Potato requires cool night temperature to induce tuberization, which is inhibited by even moderately high temperatures (Ku *et al.*, 1977; Ewing, 1997). Tuber initiation was most affected by high temperature (Ghosh *et al.*, 2000). High temperature reduces gross photosynthetic rate (Fleisher *et al.*, 2006). Although photosynthesis in potato is suppressed by high temperature (Ku *et al.*, 1977), it is not as sensitive to temperature as tuberization and partitioning of photosynthates to tubers (Reynolds *et al.*, 1990; Midmore and Prange, 1992). Therefore, even moderately high temperature drastically reduces tuber yield without much affecting the photosynthesis and total biomass production (Peet and Wolfe, 2000).

Interaction effect of temperature and CO_2: Potato tuber yield of plants exposed to high temperatures (35 °C) were extremely low regardless of CO_2 treatment, while in a non temperature stress treatment (25 °C) doubling CO_2 increased tuber yield significantly by 71.5% (Peet and Wolfe, 2000). In another study potato was grown for 35 days under CO_2 concentrations (500, 1,000, 1,500 and 2,000 micromoles mol-1) at both 16 °C and 20 °C air temperature. The mean starch concentration increased with increasing CO_2 concentration at both 16 °C and 20 °C and was consistently higher at 16 °C than at 20 °C (Cao and Tibbitts, 1997). The SLW (g.m-2) was positively related to the foliar starch concentration on the basis of leaf area or dry weight (Cao and Tibbitts, 1997).

The CO_2 enrichment does not appear to compensate for the detrimental effects of higher temperature on tuber yield, while the quality of potato is likely to be impacted severely in terms of marketable grade of tubers and internal disorders.

Quality

Elevated CO_2 increased the amount of dry matter and starch with decrease in glycoalkaloid and nitrates improving the quality of tubers (Vorne *et al.*, 2002; Schapendonk *et al.*, 2000; Donnelly *et al.*, 2001). Nearly all the nutrient elements tend to decrease in tubers under elevated CO_2 (Cao and Tibbitts, 1997; Fangmeier *et al.*, 2002) and citric acid content decreases causing a higher risk of discoloration after cooking (Vorne *et al.*, 2002).

High temperature is associated with tuber disorder of internal necrosis (Sterrett *et al.*, 1991). In a pot experiment in naturally lit glasshouse and phytotron, high temperature (30 °C) decreased the total dry matter and tuber yield and degraded quality by reducing specific gravity of tubers (Ghosh *et al.*, 2000). Nitrate reductase (NR) activity was also decreased by high temperature. The inhibition of tuber yield was due to limited translocation of carbohydrates from leaves to tubers following the reduction of NR activity and carbohydrate expense for dark respiration (Ghosh *et al.*, 2000). It can also affect tuber quality by causing 'heat sprouting' which is premature growth of stolons from immature tubers (Wolfe *et al.*, 1983; Struik *et al.*, 1989) and internal necrosis (Sterrett *et al.*, 1991). Potato processing requires large-sized tubers with high dry-matter content. Warming may reduce proportion of marketable and processing grade tubers for table and processing purposes, though dry matter may increase.

Pests of Target Crop

Effect of climate change and global warming on pests and diseases of potato is given in detail elsewhere in a separate chapter in this book (Singh and Bhat, 2010).

INDIRECT EFFECTS OF CLIMATE CHANGE AND GLOBAL WARMING

Drought: Optimal water supply is essential for potato, because of its shallow root system. The potato plant generally roots rather shallowly 40–50 cm (Beukema and Van der Zaag, 1990). Potato is extremely sensitive to drought particularly at tuber initiation with substantial loss in tuber yield. Dry matter partitioning to root, shoot, leaf and stem as a function of development stage (DS) and root:shoot ratio is affected by drought stress. Drought, while reducing dry matter production increases the root:shoot ratio indicating a shift in the balance of growth in favour of roots. Roots of plants grown in drought conditions also tend to be thinner. Both responses enable droughted plants to exploit the available soil moisture more effectively (Vos, 1995). Tuber initiation and maturity under drought stress conditions is hastened (Beukema and Van der Zaag, 1990).

***Salinity*:** Potato is highly sensitive to salinity and irrigation with saline waters with even moderate residual sodium carbonate (RSC) values (Singh and Trehan, 1993).

***Frost*:** Potato is extremely sensitive to frost. Complete loss of foliage is reported below 2 °C of ambient temperatures for 2-3 consecutive nights. More than 4-5 hours. duration of temperature below 1 °C may result in foliage loss of 50% in even one night exposure. However, yield losses depend on crop growth stage at occurrence of frost. Frosting late in the season 80–90 days after planting (DAP) results in yield loss of 10–15%, while that at 50–60 DAP may cause yield loss of 30–50%.

***Flooding*:** Flooding for even short period of 2-3 days during active vegetative phase affects growth and yield. Flooding before emergence severely affects emergence due to rotting of seed tubers and soil crust formation, while that near harvesting results in rotting and rupture of tuber lenticels affecting physical appearance and marketable quality.

***Erratic unseasonal rains*:** Rains of even 10–15 mm during planting or immediately after planting affects emergence due to soil crust formation, delaying planting operation with consequent loss in yield. Rains during active vegetative phase may promote incidence of late blight disease.

***Seed tuber production*:** In vegetatively propagated potato crop, disease-free quality seed tubers as planting material has special and added significance. Tuber as seed material is the carrier of a host of fungal, bacterial and viral diseases, responsible for rapid and drastic reduction in yield in successive generations of clonal multiplication. Most importantly, seed tubers alone accounts for half of the cost of inputs in potato cultivation and the profitability of the cropping enterprise to a large extent depends on quality of seed used. Viral diseases transmitted by aphids and other vectors are mainly responsible for rapid degeneration of planting material in potato crop. The technology of '*seed-plot technique*' was developed on the sole premise of growing seed tubers in relatively aphid free periods in plains during winters and termination of vines by dehaulming before aphid population crosses a threshold to minimize infection of viral diseases.

The appearance of potato peach aphid (*Myzus persicae*) is reported to advance by two weeks for every 1°C rise in mean temperature and population build-up is positively correlated with maximum temperature and minimum relative humidity (Biswas *et al.*, 2004; Dias *et al.*, 1980). Thus, under the impact of climate change and global warming the earlier appearance and increase in aphid population is likely to limit the aphid free period to the detriment of seed tuber quality and quantity, which will ultimately affect potato production in India. In many regions, warming may abolish seed tuber production altogether, while in others it will involve extra cost on chemicals and pesticides treatment with increased cost of seed resulting in decline in profitability.

***Storage*:** The harvesting of potato in plains of India coincides with the onset of hot summer season. Cold storage of tubers is recommended by the end of February to prevent heavy weight loss and rotting. It is cold stored up to the end of October till withdrawal for human consumption and planting is accomplished. In certain regions, relatively cooler climate permits on-farm storage of tubers for short period for 80–90 days in improvised country stores, heaps and pits with acceptable losses. However, warming from March to June may preclude this practice with increased cost of storage in the cold stores. On the other hand earlier cold storage than the recommended by the end of February and prolonged storage beyond October till weather is favourable for planting, might become necessary under global warming scenarios. This will increase the operational cost and energy use in cold stores with implied increase in cost of cold storage and enhanced market price of potato for table and seed purposes. More number of strategically located cold stores near production and consumption centers would be needed to maintain the supply chain during hot summer months from March–October under global warming.

***Potato processing industry*:** The potato tubers stored in cold stores at 4 ºC is not suitable for processing purposes due to increase in reducing sugar content imparting undesirable attributes in various processed products. Potato tubers kept in low temperature stores at 10 ºC or in on-farm country stores are utilized for processing. However, tubers kept in these low temperature and country stores suffer from sprouting and weight loss once the dormancy is broken rendering it relatively unfit for processing. Global warming will reduce the 'time window' of availability of potato suitable for processing and will result in enhanced cost of chemical treatment of tubers to prevent sprouting. This has implication for viability of potato processing industry for supply of raw material for extended periods to be economical.

REGIONAL VULNERABILITY TO CLIMATE CHANGE IN INDIA

The entire Indo-Gangetic plains, where irrigated potato is mainly grown is vulnerable. However, West Bengal with highest productivity and second largest potato-producing state in India appears highly vulnerable. Winters are mild in West Bengal and 'window' of suitable growing period is small, any rise in temperature will severely impact productivity with associated problems of storage and post-harvest handling of produce in warmer conditions. Other vulnerable states are Bihar and Uttar Pradesh, which contributes maximum in total potato production. Punjab, Haryana, and adjoining areas in northern Rajasthan and western Uttar Pradesh, where winters are relatively severe experiencing occasional frost might benefit from global warming to certain extent. The rainfed crop in plateau regions and other areas in south India would be most vulnerable due to warming and associated drought conditions.

Aberrant Weather and Extreme Events

- Rains in winter season received at planting affects emergence and delays planting with reduction in tuber yield.
- Heavy showers during the crop season resulting in flooding affects tuber yield.
- Heavy rains at the time of harvesting induce rottage in field and in temporary heaps of harvested potato in the field.
- Overcast sky and rains early in the crop season invariably increases the attack of late blight disease with severe reduction in yield.

- Early frosting received in last fortnight of December and first week of January damages the crop in North-Western plains and western Uttar Pradesh.
- Relatively warmer winters during 2008 reduced tuber yield in West Bengal, Uttar Pradesh and Bihar.

Adaptation Measures for Climate Change and Global Warming

- Use of crop residue mulches for some time after planting.
- Use drip irrigation in place of furrow and basin methods.
- Alter cultural management in potato-based cropping systems.
- Conservation tillage and on-farm crop residue management.
- Improvement and augmentation of cold storage facilities and air conditioned transportation from producing to consumption centers.
- Subsidizing additional cost of pests and water management.
- Insurance against weather for the cash crop of potato with high cost of cultivation.
- Strengthen education, research and development in warm climate production technology for ware and seed potato crop.

Mitigation Measures to Reduce Emission of CO_2/GHG and Carbon Sequestration Potential of Crops

Potato being a short duration annual crop with readily decomposable crop residues has very limited carbon sequestration potential. Other mitigation measures are as follows.

- Resource conservation techniques
- Organic farming.

FUTURE STRATEGIES FOR RESEARCH

- Quantification of regional vulnerability and impact assessment.
- Development of early warning disease forecasting systems.
- Breeding short duration and heat tolerant cultivars. Mining biodiversity to heat tolerance on priority.
- Breeding drought, salinity tolerant and disease resistant cultivars.
- Advance planning for possible relocation and identification of new areas for potato cultivation.
- Improved agronomic management for water-and fertilizer-use efficiency.
- Development of agro-techniques for warm weather cultivation and potato-based cropping systems.

CONCLUSION

Potato, a native of temperate region grown under long day conditions in mild and cool summer season in Europe and America, was introduced and adapted to tropical short day conditions in India during the last century. The growth of potato in India has been phenomenal since 1950 with increase in area, production and productivity by 6, 15 and 3 times, respectively. The crop is mainly confined to Indo-Gangetic plains in mild and cool winters in India. The autumn/winter planted crop in northern plains of India comprising Uttar Pradesh, West Bengal, Bihar, Punjab and Haryana contributes 84% of total potato production in India, where its crop is grown totally under irrigated conditions. Growth and development are affected at high temperatures encountered in the tropics. No potato crop growth is possible below 2 °C and above 30 °C. The miniumum (0–7 °C), optimum (16–25 °C) and maximum (40 °C) temperatures for net photosynthesis are reported. Potato requires cool night temperature to induce tuberization. Although photosynthesis in potato is suppressed by high temperature, it is not as sensitive to temperature as tuberization and partitioning of photosynthates to tubers.

The climate change and global warming will have a profound effect on potato growth story in India impacting every aspect of not only production and profitability, but seed multiplication, storage, marketing and processing of this perishable vegetatively propagated crop. Under the impact of future scenarios of climate change the growth projections of potato in India might be arrested or even reversed, unless effective adaptation measures are evolved for timely application and implementation. Increase in temperature and atmospheric CO_2 are interlinked occurring simultaneously under future climate change and global warming scenarios. Effect of their interaction on potato would be more relevant and of greater economic significance compared to their usually counteracting direct effects on crop growth, yield and quality. It is estimated that due to global warming potato production in India may decline by 3.16 and 13.72% from current levels by the year 2020 and 2050, respectively. The potato production will be directly affected by climate change, while there would be several indirect effects on various facets of supply, storage, utilization and acreage of the crop in future climate scenarios.

References

Allen, E.J. and Scott, R.K. 1992. The potato crop. The scientific basis for improvement. *In: Principles of Agronomy and their Appiication in the Potato Industry.* Harris, P.M. (Ed.), Chapman and Hall, London, pp. 816-81.

Benoit, G.R., Stanley, C.D., Grant, W.J. and Torrey, D.B. 1983. Potato top growth as influenced by temperatures. *American Potato Journal* **60:**489-501.

Beukema, H.P. and Van der Zaag, D.E. 1990. Introduction to potato production. Pudoc, Wageningen, 207 pp.

Bindi, M., Hacour, A., Vandermeiren, K., Craigon, J., Ojanpera, K., Sellden, G., Hogy, P., Finnan, J. and Fibbi, L. 2002. Chlorophyll concentration of potatoes grown under elevated carbon dioxide and/or ozone concentrations. *European-Journal-of-Agronomy* **17:**319-35.

Biswas, M.K., De, B.K., Nath, P.S. and Mohasin, M. 2004. Influence of different weather factors on the population build up of vectors of potato virus. *Annals of Plant Protection Sciences* **12**(2):352-55.

Burt, R.L. 1964. Influence of short periods of low temperature on tuber initiation in the potato. *European Potato Journal* **7:**197-208.

Cao, W. and Tibbitts, T.W. 1997. Starch concentration and impact on specific leaf weight and element concentrations in potato leaves under varied carbon dioxide and temperature. *Journal of Plant Nutrition* **20:**871-81.

Chen CT and Setter TL. 2003. Response of potato tuber cell division and growth to shade and elevated CO_2. *Annals of Botany* **91:**373-81.

Collins, W.B. 1976. Effect of carbon dioxide enrichment on growth of the potato plant. *Hort Science* **11:**467-69.

Craigon, J., Fangmeier, A., Jones, M., Donnelly, A., Bindi, de A., Temmerman, L., Persson, K. and Ojanpera, K. 2002. Growth and marketable-yield responses of potato to increased CO2 and ozone. *European Journal of Agronomy* **17:**273-89.

Cutter, E.G. 1992. The potato crop. The scientific basis for improvement. *In: Structure and Development of the Potato Plant.* Harris, P.M. (Ed.), Chapman and Hall, London, pp. 65-161.

Dias Jac de Souza, Yuki, V.A., Costa, A.S. and Teixeeira, P.R.M. 1980. Study of the spread of virus diseases in a warm climate as compared to a cold climate, with a view to obtaining seed potatoes with a low rate of virus diseases. *Summa Phytopathologica* **14:**58-59.

Donnelly, A., Lawson, T., Craigon, J., Black, C.R., Colls, J.J. and Landon, G. 2001. Effects of elevated CO_2 and O_3 on tuber quality in potato (*Solanum tuberosum* L.). *Agric. Ecosyst. Environ.* **87:**273-85.

Donnelly, A., Craigon, J., Black, C.R., Colls, J.J. and Landon, G. 2001. Elevated CO_2 increases biomass and tuber yield in potato even at high ozone concentrations. *New-Phytologist* **149:**265-74.

Ewing, E.E. 1997. Potato. In Wien H.C. (ed.) *The Physiology of Vegetable crops.* CAB International, Wallingford, UK, pp. 295-344.

Fangmeier, A., Temmerman de L., Black, C., Persson, K. and Vorne, V. 2002. Effects of elevated CO_2 and/or ozone on nutrient concentrations and nutrient uptake of potatoes. *European Journal of Agronomy* **17:**353-68.

Finnan, J.M., Donnelly, A., Burke, J.I. and Jones, M.B. 2002. The effects of elevated concentrations of carbon dioxide and ozone on potato (*Solanum tuberosum* L.) yield. *Agric Ecosyst Environ.* **88:**11-22.

Fleisher, D.H, Timlin, D.J. and Reddy. V.R. 2006. Temperature influence on potato leaf and branch distribution and on canopy photosynthetic rate. *Agronomy Journal* **98:**1442-52.

Ghosh, S.C., Asanuma, K., Kusutani, A. and Toyota, M. 2000. Effect of temperature at different growth stages on nonstructural carbohydrate, nitrate reductase activity and yield of potato (*Solanum tuberosum*). *Environment Control in Biology* (Japan) **38:**197-206.

Heagle, A.S., Miller, J.E. and Pursley, W.A. 2003. Growth and yield responses of potato to mixtures of carbon dioxide and ozone. *Journal of Environmental Quality* **32:**1603-10.

IPCC. 2007. Climate change 2007:Fourth assessment report of the Intergovernmental Panel on climate change (IPCC), WMO, UNEP.

Katny, M.A.C., Hoffmann Thoma, G., Schrier, A.A., Fangmeier, A., Jager, H.J. and Bel van A.J.E. 2005. Increase of photosynthesis and starch in potato under elevated CO2 is dependent on leaf age. *Journal of Plant Physiology* **162:**429-38.

Khan, I.A., Deadman, M.L., Al Nabhani, H.S. and Al Habsi, K.A. 2003. Interactions between temperature and yield components in exotic potato cultivars grown in Oman. *Acta Horticulturae* **619:**353-59.

Kooman, P.L. and Haverkort, A.J. 1995. Modelling development and growth of the potato crop influenced by temperature and daylength: Lintul-Potato. *In: Potato Ecology and Modelling of Crops under Conditions Limiting Growth.* Haverkort, A.J. and MacKerron, D.K.L. (Eds), Kluwer Academic Publishers, Dordrecht, pp. 41-60.

Ku G., Edwards, E. and Tanner, C.B. 1977. Effects of light, carbon dioxide and temperature on photosynthesis, oxygen inhibition of photosynthesis and transpiration in *Solanum tuberosum. Plant Physiology* **59:**868-72.

Lawson, T., Craigon, J., Collis, J.J., Landon, G. and Weyers, J.D.B. 2002. Impact of elevated CO_2 and O_3 on gas exchange parameters and epidermal characteristics in potato (*Solanum tuberosum* L.). *Journal of Experimental Botany* **53:**737-746.

Lawson, T., Craigon, J., Tulloch, A.M., Black, C.R., Colls, J.J. and Landon, G. 2001. Photosynthetic responses to elevated CO2 and O3 in field-grown potato (*Solanum tuberosum*). *Journal of Plant Physiol.* **158:**309-23.

Magliulo, V., Bindi, M. and Rana, G. 2003. Water use of irrigated potato (*Solanum tuberosum* L.) grown under free air carbon dioxide enrichment in central Italy. *Agriculture, Ecosystems and Environment* **97:**65-80.

Marwaha, R.S. and Sandhu, S.K. 2002. Yield, growth components and processing quality of potatoes as influenced by crop maturity under short and long days. *Advances in Horticultural Sciences* **16:**47-52.

Midmore, D.J. and Prange, R.K. 1992. Growth responses of two *Solanum* species to contrasting temperatures and irradiance levels: relations to photosynthesis, dark respiration and chlorophyll fluorescence. *Annals of Botany* **69:**13-20.

Miglietta, F., Magliulo, V., Bindi, M., Cerio, L., Vaccari, F.P., Loduca, V. and Peressotti, A. 1998. Free air CO2 enrichment of potato (*Solanum tuberosum* L.): development, growth and yield. *Global Change Biology* **4:**163-72.

Olivo, N., Martinez, C.A. and Oliva, M.A. 2002. The photosynthetic response to elevated CO2 in high altitude potato species (*Solanum curtilobum*).*Photosynthetica* **40:**309-13.

Peet, M.M. and Wolfe, D.W. 2000. Crop ecosystem responses to climate change: Vegetable crops. *In: Climate Change and Global* Crop Production. Reddy, K.R. and Hodges, H.F. (Eds). CAB International, pp. 213-43.

Reynolds, M.P., Ewing, E.E. and Owens, T.G. 1990. Photosynthesis at high temperature in tuber bearing *Solanum species. Plant Physiology* **93:**791-97.

Schapendonk, A.H.C.M., Oijen van, N., Dijkstra, P., Pot, C.S., Jordi, W.J.R.M. and Stoopen, G.M. 2000. Effects of elevated CO_2 concentration on photosynthetic acclimation and productivity of two potato cultivars grown in open-top chambers. *Australian Journal of Plant Physiology* **27:**1119-30.

Scott ,G.J. 2002. Maps, models and muddles: World trends and patterns in potatoes revisited. *Potato Research* **45:**45-77.

Singh, J.P. and Trehan, S.P. 1993. A case study of soil and irrigation water related constraints in potato crop production: Cause and correction. *National Seminar on 'Developments in Soil Science*, 8-12, October held at Dehradun, pp. 185-86.

Singh, J.P., Govindaskrishnan, P.M., Lal, S.S. and Aggarwal, P.K. 2005. Increasing the efficiency of agronomy experiments in potato using INFOCROP-POTATO model. *Potato Research* **48:**131-52.

Singh, J.P., Govindakrishnan, P.M., Lal, S.S. and Aggarwal, P.K. 2008. Infocrop-Potato a Model for Simulating Growth and Yield of Potato in the Sub-Tropics. Central Potato Research Institute, Shimla, *Technical Bulletin* No. **87:**1-30.

Singh, J.P., Lal, S.S. and Pandey, S.K. 2009. Effect of climate change on potato production in India. *Central Potato Research Institute, Shimla, Newsletter* **40:**17-18.

Sterrett, S.B., Lee, G.S., Henninger, M.R. and Lentner, M. 1991. Predictive model for onset and development of internal heat necrosis of 'Atlantic' potato. *Journal of the American Society for Horticulture Science* (USA). **116:**701-705.

Struik, P.C., Geertsema, J. and Custers, C.H.M.G. 1989. Effects of shoot, root and stolon temperatureon the development of potato plant. III. Development of tubers. *Potato Research* **32:**151-58.

Vaccari, F.P., Miglietta, F., Giuntoli, A., Magliulo, V., Cerio, L. and Bindi, M. 2001. Free air CO2 enrichment of potato (*Solanum tuberosum* L.). Photosynthetic capacity of leaves. *Italian Journal of Agronomy* **5:**3-10.

Van de Geijn, S.C. and Dijkstra, P. 1995. Physiological effects of changes in atmospheric carbon dioxide concentration and temperature on growth and water relations of crop plants. *In: Potato Ecology and Modelling of Crops under Conditions Limiting Growth.* Haverkort, A.J. and MacKerron, D.K.L. (Eds). Kluwer Academic Publishers, Dordrecht, pp. 89-100.

Van Keulen, H. and Stol, W. 1995. Agro-ecological zonation for potato production. *In: Potato Ecology and Modelling of Crops under Conditions Limiting Growth.* Haverkort, A.J. and MacKerron, D.K.L. (Eds.), Kluwer Academic Publishers, Dordrecht, pp. 357-72.

Vandermeiren, K., Black, C., Lawson, T., Casanova, M.A. and Ojanpera, K. 2002. Photosynthetic and stomatal responses of potatoes grown under elevated CO2 and/or O3 - results from the European CHIP-programme. *European Journal of Agronomy.* **17:**337-52.

Vorne, V., K. Ojanpera, de L. Temmerman, M. Bindi, P. Hogy, M. B. Jones, T. Lawson and K. Persson. 2002. Effects of elevated carbon dioxide and ozone on potato tuber quality in the European multiple-site experiment 'CHIP-project'. *European-Journal-of-Agronomy* **17:**369-81.

Vos, J. 1995. Nitrogen and the growth of potato crops. *In: Potato Ecology and Modelling of Crops under Conditions Limiting Growth.* Haverkort, A.J. and MacKerron, D.K.L. (Eds), Kluwer Academic Publishers, Dordrecht, pp. 115-28.

Wheeler, R.M., Tibbitts, T.W. and Fitzpatrick, A.H. 1991. Carbon dioxide effects on potato growth under different photoperiods and irradiance. *Crop-science* (USA) **31**(5): 1209-13.

Wolfe, D.W., Fereres, E. and Voss, R.E. 1983. Growth and yield response of two potato cultivars to various levels of applied water. *Irrigation Science* **3:**211-22.

13

Impact of Climate Change on Onion and Garlic Production

K. E. Lawande

CLIMATE and weather play a predominant role in crop growth and productivity. Environmental stress is the primary cause of crop losses worldwide, reducing average yield for most of the crops by more than 50% (Boyer *et al.*, 1982; Bray *et al.*, 2000). Aberrant weather conditions have always been unpredictable and often impairing realizable crop yields. Sporadic heat or cold wave conditions and extensive variability in quantity and distribution of rainfall have been causing severe crop damages and extensive economic losses to farmers. Now, it is widely acknowledged worldwide that climate is changing at a faster rate. Global warming mainly due to increased emissions of greenhouse gases (GHG) from industrial and other technological advances of modern world is responsible for climate change. The concentration of GHG's carbon dioxide, methane and nitrous oxide is increasing at alarming rate. Temperature variations more than critical levels are experienced. Total rainfall might not have changed but the distribution has changed largely. Climate change and global warming are most likely to influence the severity of environmental stress imposed on vegetable crops.

IMPACT ANALYSIS ON ONION AND GARLIC

The onion and garlic crops are cool season crops and perform well during winter followed by early part of summer season. In onion and garlic, bulb initiation takes place between 10 and15 °C night and 20 and 25 °C day temperatures. Bulb development is best at 18–20 °C nights and 25–30 °C day temperatures. For maturity, day temperatures of 35–38 °C is required. High night temperatures at bulb initiation stage lead to poor bulb development. Very low (<10 °C) temperatures during bulb development lead to bolting of onion and clove sprouting in garlic. Excessive high temperatures (>42 °C) at the time of maturity during April-May lead to reduction in size of bulbs and such bulbs do not store well for longer time. Besides, temperature, optimum humidity, bright sunshine and moderate rainfall play great role in productivity. Bright sunshine of 12-13 hours during bulb development and maturity, humidity between 65–75 and 500–600 mm and well distributed rainfall ensures good crop growth.

Elevated CO_2 and Temperature

Productivity

The information available on effects of climate change including the interaction of carbon dioxide and temperature, on growth, development and yield of vegetable crops is scanty. Generally, crops show a positive effect of carbon dioxide on yield. Elevated carbon dioxide concentration from 325 to 530 ppm increased yield of root crops by 34%, which is larger and more positive than most other crops (Kimball, 1983). In onion, elevated carbon dioxide stimulates growth with no overall effect on crop duration (Wheeler *et al.*, 1996; Daymond *et al.*, 1997; Wurr *et al.*, 1998). An increase in carbon dioxide levels, from current ambient levels to 530 ppm, lead to an increase in bulb onion yield of 30–50%.

Crops that are harvested later in their physiological development or at maturity like onion and cauliflower have shown a negative effect of increasing temperature. In onion, warmer temperatures shorten the duration (Wheeler *et al.*, 1996; Daymond *et al.*, 1997; Wurr *et al.*, 1998). In general, it appears that a 1°C increase in average temperature decreases bulb yield by 3.5–15%.

The net effect of increased carbon dioxide levels and increased temperatures depends on their interaction and varies with species. Further, combined with elevated temperatures, decreased precipitation could cause reduction in irrigation water availability and may increase evapo-transpiration leading to severe crop water stress condition (IPCC, 2001). The problems from climate change also arise from extreme events that are difficult to predict. Erratic rainfall patterns and unpredictable high temperature spells consequently reduce crop productivity. A case study of Pune for last 10 years indicated that rainfall and maximum temperatures varied greatly than normal range.

Excessive rainfall, cloudy weather in *kharif* and late *kharif* season lead to crop failures up to 60–70%. The soil-borne and fungal disease build-up is very high during *kharif* season.

Quality

High rainfall in August September spoiled onion, while high temperatures during March and April reduced yield and storage quality of onion and garlic. Very low temperatures (<10 °C) for more than one week in January-February caused high percentage of bolting in onion and premature sprouting of garlic cloves due to breaking of dormancy.

Pests of Target Crop

Onion thrips is the major pest and causes considerable damage to crop, further it transmits the diseases too. Study conducted on seasonal incidence of onion thrips at National Research Centre for Onion and Garlic, Rajgurunagar, showed that thrips population reached to maximum (48/plant) in August during *kharif* season and touched the lowest (1/plant) in October (NRCOG, 2001). Thrips started multiplying from December and reached highest (112/plant) in February and drastically came down to 14/plant during March. During the entire onion growing period, two peaks of thrips population were recorded one in August and the second in February. This confirmed that low humidity during this period favours the thrips incidence. Incidence of disease is more related with temperature and humidity. Higher temperature with higher humidity enhances the disease. Excess rainfall associated with poor drainage enhances soil-borne diseases.

Aberrant Weather and Extreme Events

- The photo-thermo sensitive crops like onion and garlic are affected extensively by the climatic aberrations. Excessive rainfall experienced in the year 1998 and 2005 caused heavy crop losses. High temperatures in earlier parts of the crop seasons followed by heavy rainfall during these years drastically reduced yield, resulting in short supply in domestic market.
- In case of *rabi* season, sudden rise of temperature during bulb development stage often reduce the size of bulb and affect storage quality also.

Adaptation Measures for Climate Change and Global Warming

- Agronomic management of date of planting to escape periods of high rainfall and temperature.
- Planting on broad bed furrow (BBF) in stead of traditional flat beds may reduce losses from flooding in *kharif* season and salinity in winter-summer season.
- BBF planting coupled with drip irrigation can help overcome drought and salinity conditions.
- Irrigation by micro-sprinklers can avert the micro temperature in summer season by reducing effect of high temperature on bulb development.
- Integrated pest management systems, which adjust control measures to the observed problem and also take a range of influencing factors (including weather) into account to cut cost on pesticides.
- Conservation tillage (the practice of leaving some of the previous season's crop residues on the soil surface) and mulching.
- Introduction of water saving techniques such as drip irrigation coupled with fertigation. Recent studies showed savings of 40% water, 30% labour, 30% nitrogen with an yield increase of 15% by drip irrigation (NRCOG, 2003).
- Mulching with organic waste and bicolour polythene not only conserves moisture but control weed population and maintains soil temperature.

Mitigation Measures to Reduce Emission of CO_2/GHG and Carbon Sequestration Potential of Crop

Onion and garlic being annual crops have hardly any significant carbon sequestration potential. Other mitigation measures are as follows:

- Resource conservation techniques
- Organic farming.

FUTURE STRATEGIES FOR RESEARCH

- Development of heat, salinity and drought tolerant varieties with pest and disease resistance.

- Developments of varieties for late *kharif* season with long dormancy and tolerance to flooding to avoid direct bolting in onion and premature sprouting of garlic cloves.

Development of infrastructure for improved varieties through mass selection, QTL mapping and incorporation of genes resistant to abiotic stress employing conventional as well as biotechnological approaches.

CONCLUSION

The onion and garlic crops are cool season crops that perform well during winter followed by early part of summer season. Climate is changing at a faster rate. Climatic aberrations always affect the photo-thermo sensitive crops such as onion and garlic badly. Excessive high and low temperatures affect bulb development and their size. High rainfall and cloudy weather aggravates diseases. Temperature rise of 1 °C over the normal reduces onion yield by 34%. Increased level of CO_2 increases yield by 30–50% under normal range of temperatures. High rainfall, cloudy weather and ill drained conditions reduce yield by 50–90% and conditions resulting high salinity reduce size of bulbs. Developing varieties resistant/tolerant to aberrant weather conditions would be crucial in future. Innovative agronomic practices, viz. raised bed planting, micro-irrigation, fertigation, mulching *etc*. are possible adaptation measures to climate change in onion and garlic.

References

Boyer, J.S. 1982. Plant productivity and environment. *Science* **218**:443-48.

Bray, E.A., Bailey-Schnell, R.C., Higgins, R.W., Lowrimore, J., Kousky, V.E., Tinker, R., Thiaw, W., Chelliah, M. and Artusa, A. 2000. Climate assessment for 1999. Supplement June 2000. *Bulletin of American Meteorology Society*. Vol. **81.**

Daymond, A.J., Wheeler, T.R., Hadley, P., Ellis, R.H. and Morison, J.I.L. 1997. Effects of temperature, CO2 and their interaction on the growth, development and yield of two varieties of onion (*Allium cepa* L.). *Journal of Experimental Botany* **30**:108-18.

IPCC. 2001. Climate change 2001. Impacts, adaptation and vulnerability, Intergovernmental Panel on Climate Change, New York, USA.

Kimball, B.A. 1983. Carbon dioxide and agricultural yield: An assemblage and analysis of 430 prior observations. *Agronomy Journal* **75**:779-88.

NRCOG. 2001. Annual report-2000-01. National Research Centre for Onion and Garlic, Rajgurunagar, Pune, Maharashtra. pp.57.

NRCOG. 2003. Annual report-2002-03. National Research Centre for Onion and Garlic, Rajgurunagar, Pune, Maharashtra, pp.34-35.

Wheeler, T.R., Ellis, R.H., Hadley, P., Morison, J.I.L., Batts, G.R. and Daymond, A.J. 1996. Assessing the effects of climate change on field crop production. *Aspects Appl. Biol.* **45:**49-54.

Wurr, D.C.E., Hand, D.W., Edmondson, R.N., Fellows, J.R., Hannah, M.A. and Cribb, D.M. 1998. Climate change: a response surface study of the effects of CO_2 and temperature on the growth of beetroot, carrots and onions. *Journal of Agriculture Science, Camb.* **131:**125-33.

14

Impact of Climate Change on Cassava and Sweet Potato

V. Ravi and C. S. Ravindran

Continued build-up of atmospheric CO_2 and other greenhouse gases like methane (CH_4), nitrous oxide (N_2O) and CFCs is attributed for rise in temperature of earth surface. The mean global surface air temperatures may increase by 3–5 °C from present levels. Heat stress is one of the most important factors affecting crop yield (Wrigley *et al.*, 1994). All plant processes are sensitive to and can be irreversibly damaged by heat. Elevated temperatures accelerate senescence, reduce the duration of viable leaf and diminish photosynthesis (Harding *et al.*, 1990).

Cassava is an important crop in Tamil Nadu, Eastern Andhra Pradesh and Kerala. Cassava finds utilization in three different sectors, viz., human consumption, animal feed and industries in production of sago (*sabudhana*), starch, and other value-added products like dextrin, liquid glucose, adhesives, etc. (Balagopalan, 2000). Sweet potato is an alternative food crop due to its high production potential, high nutritive value and ability to survive in natural disaster. The kind of climate scenario being projected for the middle of the 21st century may lead to shifts from drought susceptible crops to more drought/flood tolerant crops like cassava, sweet potato, sorghum and pearlmillet. Thus impact of climate change on tropical tuber crops like cassava and sweet potato assumes importance for food security in tropical India.

IMPACT ANALYSIS ON CASSAVA/SWEET POTATO

Monsoon is Crucial for Cassava

In Tamil Nadu, cassava is grown under partially irrigated conditions or as a rainfed crop in areas where the annual rainfall varies from 900 to 1500 mm. Under rainfed conditions, cassava is planted in November-December and harvested after 10-11 months. The crop is exposed to higher day temperatures (35–40 °C) during March—May along with water deficit stress causing significant reduction in tuber/starch yield. In Eastern Andhra Pradesh (East Godhavari District), cassava is planted in June (with the onset of south-west monsoon) and the last 3-4 months of its growth period coincides with a seasonal drought (December-May) causing significant reduction in tuber yield. In both Tamil Nadu and Andhra Pradesh, contingent drought due to failure of rain during monsoon is attributed for reduction in tuber yield (Fig. 14.1).

Mild water deficit stress (WDS) is favourable for tuber growth. However, under prolonged WDS conditions (WDS >20 days), both vegetative growth as well as tuber bulking ceases, plant sheds about 80% of leaves and the plant becomes dormant. When conditions become favourable with adequate soil moisture the plant resurrects and resumes its growth, develops canopy consuming considerable amount of starch from tubers and tuber continues to bulk. Reduction in photosynthetic rate, total dry matter production and tuber yield (28–42%) under WDS conditions has been reported (Ramanujam, 1990). However cassava was found to maintain nearly 50% of its photosynthetic rate under drought conditions (Ravi, 2000) offering a better choice of crop in future warm climatic conditions. Wide genetic diversity exists among genotypes about tolerance to drought (Ramanujam, 1990; Ravi and James George, 2003).

Elevated CO_2 and Temperature on Cassava

The effect of elevated CO_2 and high temperature on growth and productivity, particularly starch yield of cassava is yet to be investigated. Literature review showed that information available on the response of cassava to elevated CO_2 is limited except the works by Imai *et al.*, (1984) in Japan under controlled conditions.

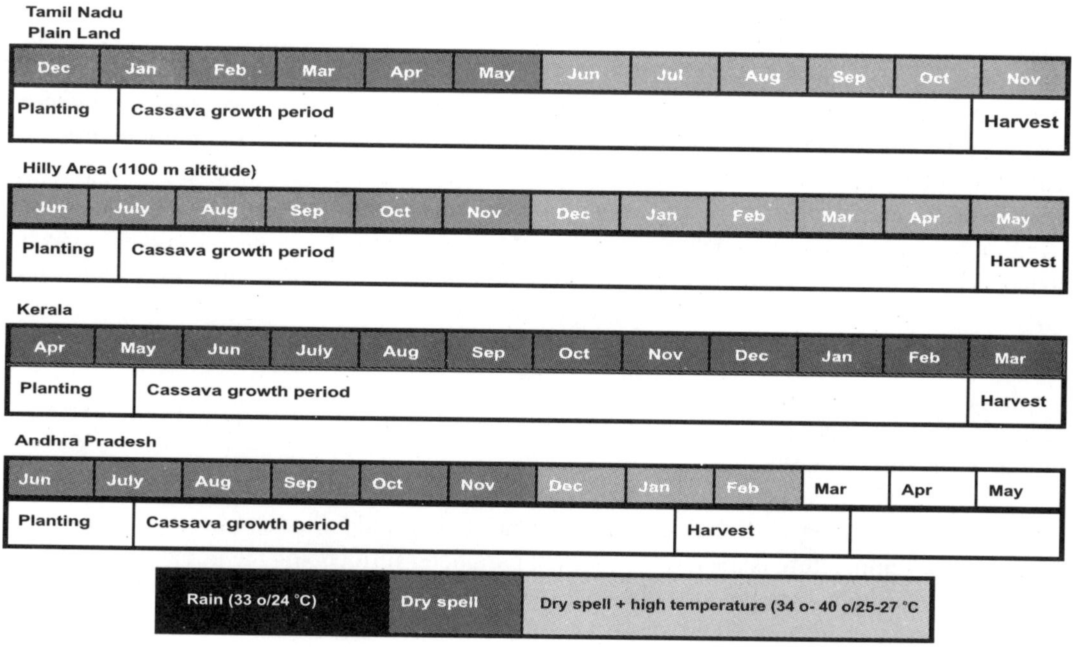

Fig. 14.1. Production situations: cropping season of cassava in 3 major states

Productivity

Net photosynthesis increased in cassava up to 500 ppm of elevated CO_2 with negligible increase thereafter (Mahon *et al.*, 1977). Combination of high CO_2 (700 ppm) and high temperature (33/26 °C day/night temperature) increased the tuber yield over control plants grown at 350 ppm CO_2 (Mahon *et al.*, 1977; Imai *et al.*,1984).

Photosynthetic rate of cassava leaves has been reported to be optimum (90–100%) between 30 and 35 °C maximizing at 25 °C (El-Sharkawy and Cock 1990; El-Sharkawy *et al.*, 1992; Mahon *et al.*, 1977). Field experiments showed that leaves of cassava plants grown at 29/24 °C day/night temperature regime had higher photosynthetic rate (19.0 mg CO_2 $dm^{-2}h^{-1}$) than plants grown at 24/19 °C (12.0 mg CO_2 $dm^{-2}h^{-1}$). The rate of increase in photosynthesis was 1.4 mg CO_2 $dm^{-2}h^{-1}$ per 1 °C rise in temperature (Mahon *et al.*, 1977). However, yield response to rise in temperature varies among genotypes (Irikura *et al.*, 1979).

Cassava being a C_3 plant has a unique, partial Krant'z anatomical feature in leaf (Edwards *et al.*, 1990; Angelov *et al.*, 1993; Ueno and Agarie, 1997; Clatayud *et al.*, 2002). It has 15–25% phosphoenol pyruvate corboxylase (PEP case) activity of maize and sorghum. The PEP case activity of cassava is greater than typical C_3 plants and equivalent to that of C_3–C_4 intermediates. The PEP case activity of cassava increased by 13%, whereas the RUBPcase activity decreased by 42% under water deficit stress and heat stress conditions (El-Sharkawy, 2006). Thus, cassava possesses a greater advantage to respond to elevated CO_2 concentrations.

Quality

Cassava may sustain vegetative growth and biomass at high temperatures (40 °C) under adequate soil moisture, but sucrose synthesis and export from leaves and starch synthesis in tubers are affected. The enzymes involved in starch biosynthesis are highly sensitive to higher atmospheric temperatures (>30 °C). The effect of temperature on regulation of starch synthesis in cassava is yet to be investigated in detail. However, greater starch content by 3–5% in cassava tubers is reported at cooler climate in high altitude than warmer conditions in plain. Therefore, effects of elevated CO_2 on cassava tuber yield and starch appear to be temperature dependent. Higher temperatures and drought also can divert photosynthates to lignification (Amthore, 2003) and this metabolic change needs detailed investigation as it drastically reduces starch yield. Thus, high temperatures (>30°C) and drought can significantly reduce the tuber yield and starch content of cassava/sweet potato tubers.

Pests of Target Crop Cassava

In cassava, mites and spiral whitefly are major pest problems causing serious reduction in crop growth and tuber yield, whereas white fly (*Bemisia tabaci*) acts as the vector of cassava mosaic virus spreading the disease.

The whitefly (*Bemisia tabaci* (Genn.) accounts for 4–30% secondary spread of the Indian cassava mosaic disease (ICMD) depending upon the variety, under field conditions (Palaniswami *et al.*, 1996). Temperature is the most important parameter influencing the population build-up of whitefly and high population build-up is recorded during February-March. The whitefly population and temperature significantly correlated (r = 0.86), whereas rainfall and RH were not correlated (Pillai and Daniel, 1979).

Tetranychid spidermites are the most serious pests of cassava. Four species of mites, *viz.*,

Tetranychus neocaledonichus, *T. telarius*, *Eutetranychus orientalis* and *Oligonychus biharensis* are the spidermite complex in cassava causing severe reduction in tuber yield depending upon the season. Reduction in tuber yield varies from 17 to 23% during July–April (south-west monsoon) season and between 20 and 30% during November–August (north-east monsoon) season due to spidermite infestation. Under severe mite infestation, yield loss from 49 to 57% was observed (Pillai and Palaniswami 1990). The spidermite population is very low from July–December (1–17 mites/leaf) and thereafter the build-up increased. The peak periods are from January to April (193-430 mites/leaf) when the rainfall and RH are low and temperature is the highest. A fall in RH below 75% and an increase in maximum temperature >31 °C are highly favourable and conducive for rapid multiplication of mites. Rainfall is the most important limiting factor as the population suddenly decreases during May with the onset of monsoon. Heavy rains accompanied with strong winds dislodge eggs, nymphs and adults of mite complex. The residual population remaining after the rain multiply rapidly and become epidemic. An outbreak of spidermites can be expected every year from January to April if there is no heavy rain during the period.

The incidence of spiralling whitefly (*Aleurodicus disperses*) on cassava was first reported in 1993-94 in Kerala (Palaniswami *et al*., 1995). It is as a major pest on cassava in Kerala and Tamil Nadu. In severe infestation, the leaves crinkle, curl and shed. Incidence of spiralling whitefly was observed throughout the year. The population was high during March-April. In Thiruvananthapuram area, mean infestation varied from 25 to 62% and mean population varied from 94 to 400/leaf. In Tamil Nadu, mean infestation varied from 16 to 54% and mean population 10 to 425/leaf. The yield loss due to infestation by spiral whitefly may vary from 22.1 to 26.15%.

Diseases of Cassava

The cassava mosaic virus disease (CMD) and tuber rotting are major diseases affecting tuber yield and quality of cassava. In vegetatively propagated cassava crop, the CMD primarily spreads through infected planting material (stems) and later through the insect vector *Bemisia tabaci*. Losses in tuber yield due to CMD ranges from 10 to 90% with mean of 30% (Edison, 1979). In susceptible varieties (H-226, H-165) 100% infection was at observed in farmers' field in Tamil Nadu and Kerala. Cassava Sree Visakham M-4 and Sree Padmanabha are tolerant to CMD. In cool climate and higher altitudes (1100 m), CMD incidence and severity is less than plains.

Tuber rottage caused by soil born *Phytophthora palmivora* may result in losses from 50 to 100% of tuber yield and quality in Tamil Nadu. High rainfall, waterlogging, heavy soil and poor drainage are attributed for appearance of the disease. Survival of P. *palmivora* is 24 weeks at 40% soil moisture, but only 6 weeks in air dry soil.

Elevated CO_2 and Temperature on Sweet Potato

The information available on the effect of elevated CO_2 and temperature on sweet potato is scanty based on works mostly done abroad. Therefore, it is important to investigate under Indian conditions and varieties the physiology, metabolism, growth, regulation of starch biosynthesis of cassava/sweet potato in relation with high temperature (>30 °C) and elevated CO_2 conditions.

Elevated CO_2 of 514 and 666 ppm increased the photosynthetic rate, specific leaf weight (SLW), biomass and tuber yield of sweet potato (Bhattacharya *et al.*, 1990 b,1992; Biswas *et al.*, 1986). The tuber yield at 363, 514 and 666 ppm of CO_2 was 346.9, 388.2 and 395.9 g/plant corresponding to 28.91, 32.35 and 32.99 tonnes/ha (Bhattacharya *et al.*, 1992). It also ameliorated the effect of drought with increase in Root:shoot ratio (Bhattacharya *et al.*, 1990 a).

Tuber yields of sweet potato are usually greater in winter (*rabi*) season than in warm rainy (*kharif*) season. The effect of temperature rise on tuber yield of sweet potato was simulated using SPOTCOMS model (Santhosh Mithra and Somasundaram, 2008). The results showed variable effect depending on temperature range. Rise in temperature by 1 °C from 19.24–20.24 °C increased the tuber yield by 1.26 tonnes/ha, whereas that from 28.24–29.24 decreased the yield by 0.12 tonnes/ha (Table 14.1). The tuber yield decreased by 2.37 tonnes/ha due to 1 °C rise in temperature from 23.5–25.9 °C (Nawale and Salvi, 1983).

Enzymes of photosynthesis (rubisco activase) as well as electron transport in sweet potato leaves are sensitive to high temperature >35 °C (Cen and Sage, 2005). Photosynthetic rate of sweet potato leaves increased with increase in temperature up to 34 °C and increase in CO_2 up to 560 ppm. At temperature greater than 34 °C, increase in CO_2 did not increase photosynthetic rate.

Table 14.1. Simulation of effect of 1 °C rise in temperature on sweet potato yield by SPOTCOMS model.

Variety	*Seasonal mean temperature (°C)*		*Seasonal mean temperature (°C)*	
	19.24	20.24	28.24	29.94
Sree Arun	5.01	6.23	8.2	8.36
Sree Bhadra	4.17	6.45	7.76	7.49
Sree Rethna	7.97	8.25	10.24	9.98
Mean	5.72	6.98	8.73	8.61

Source: Santhosh Mithra and Somasundaram (2008).

Quality

Carotene, starch and glucose content in sweet potato tuber increased under high CO_2 (675 ppm) (Bhattacharya *et al.*, 1989, 1990a).

Pests of Target Crop Sweet Potato

Sweet potato weevil (*Cylas formicarius*) is the major pest causing tuber yield losses up to 79% and quality at farmers' fields (Pillai 1994). In Kerala, tuber damage was greater in upland (4-50%) than in lowland (0-22%) conditions. The tuber damage depends on season, temperature, rainfall, soil moisture, method of planting, soil type, time of harvesting and also cropping pattern (Palaniswami *et al.*, 1990). Both tuber damage and weevil population build-up start between second and third month of the crop and increases rapidly with age of the crop. Experiments showed that the mean weevil population and per cent of tuber damage were the lowest during June–September (*kharif*) season than during February–May (summer) season (Table 14.2). Tuber damage was comparatively lower during rainy season/irrigated conditions and higher during summer season/rainfed conditions (Palaniswami, 1990). Higher temperature favours the pest. The developmental period of sweet potato weevil took 69–80 days at 22 ± 1 °C as compared to 26 to 35 days at 29 ± 2 °C (Rajamma and Pillai, 1987). High soil moisture restricts easy access to tuber by weevil. Rainy season planting and irrigation to maintain high moisture content in soil effectively minimizes the attack of sweet potato weevil.

Table 14.2. Sweet potato tuber damage as influenced by rainfall

Season	*Mean temperature (°C)*			*Mean RH (%)*	*Total rainfal (mm)*	*Mean weevil population per plant*	*Tuber damaged (%)*
	Maximum	Minimum	Mean				
June — September	29.45	23.33	26.39	84.88	1161.7	0.74	44.75
Feb — May	32.58	24.43	28.51	73.0	368.4	1.07	71.0

Source: (Rajamma and Padmaja, 1981)

Aberrant Weather and Extreme Events

The tuber yield of cassava as affected by annual mean temperature of Salem in Tamil Nadu during 20 years period from 1986 to 2005 is given (Fig. 14.2). The mean annual temperature in Salem, which is cassava-growing area in Tamil Nadu, varied between 27.7 to 28.9 °C with a mean annual temperature of 28.3 °C over 20 years period. The mean tuber yield during this period increased from 29.3 tonnes/ha (1986–91) to 36.8 tonnes/ha (1991–2005). However, during this period the annual mean temperature remained below 30 °C, increase in annual mean temperature exceeding 30 °C may decrease tuber yield.

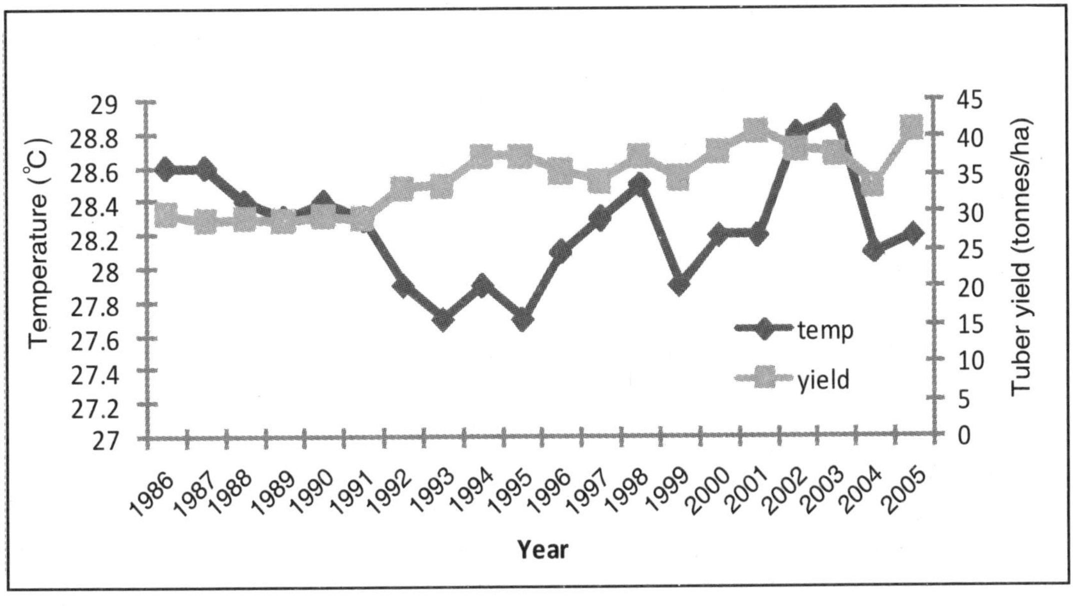

Fig. 14.2. Tuber yield versus temperature in Salem, Tamil Nadu during 1986–2005

Mitigation Measures to Reduce Emission of CO_2/GHG and Carbon Sequestration Potential of Crops

The carbon sequestration studies are yet to be conducted in cassava and sweet potato. Therefore, information is not available. Other mitigation measures are as follows:

- Resource conservation techniques
- Organic farming.

Adaptation Measures for Climate Change and Global Warming

- Agronomic management of planting dates, fertilization and irrigation.
- Irrigation during critical period of crop growth and conservation of soil moisture.
- Raised bed cultivation to avoid flooding and save water.
- Intensifying multiple cropping systems.
- Use of biopesticides, and organic formulations ("Panchakavya"/ "Dasakavya") for pest control and increased crop productivity.
- Exploring availability of traditional, indigenous technologies/knowledge, crop varieties and evaluating them for possible exploitation under climate change.

FUTURE STRATEGIES FOR RESEARCH

- Development of hybrids and transgenic varieties for heat, drought and pest tolerance.
- Evaluation of existing high-yielding varieties under elevated CO_2 and temperature (>35 °C) for heat and drought tolerance.
- Basic research on the effect of high temperature (35–40 °C) on tuber yield, starch/sucrose metabolizing enzymes during tuberization in high yielding cassava/sweet potato varieties.
- Identify the threshold level of temperature and CO_2 for cassava and sweet potato.
- Understand the interaction between carbohydrate metabolism and N nutrition.
- Studies on the role of calcium and oxidative stress metabolism in imparting heat /drought tolerance.
- Studies on changes in leaf proteomic (heat shok proteins, HSPs, small HSps) and transcriptomic (mRNAs, micro RNAs) profile.
- Isolation of cDNAs of genes expressed or suppressed during tuberization under high temperature/drought.
- Studies on dynamics of pests and diseases in cassava and sweet potato under climate change scenarios.
- Studies on Soil C pools and dynamics in relation to below ground carbon sequestration.

CONCLUSION

Cassava and sweet potato are important food crops for people living in remote, rural villages, hilly regions and in coastal areas. They are also important industrial crops yielding a large number of value-added products. Monsoon is crucial for cassava. Drought due to failure of rain during monsoon is mainly attributed for reduction in tuber yield. Research conducted abroad showed that cassava possess greater advantage to respond to high CO_2 as it contains

15–25% greater PEP case activity than typical C_3 plants and equal to C_3-C_4 intermediate crops. Similarly in sweet potato, photosynthetic rate of leaves increased between 20 and 34 °C and CO_2 from 200–560 ppm. Simulation studies with SPOTCOMS model showed increase in tuber yield by 1.26 tonnes/ha due to 1 °C rise in temperature between 19.24 and 20.24 °C, whereas increase in tuber yield was insignificant due to 1 °C rise in temperature between 28.24 and 29.24 °C. Studies on response of sweet potato to high CO_2 temperature (35–40 °C) is lacking.

In cassava whitefly a vector of cassava mosaic virus disease (CMD), spider mites and spiralling whitefly are major pests and tuber rotting is the major disease causing severe reduction of tuber yield in Tamil Nadu and Kerala. Temperature strongly influences the build-up of whitefly population. Rainfall is the most important factor influencing build-up of spidermite population. High rainfall (200–400 mm/month) decreased the mite population, while low rainfall coupled with high temperature (31 °C) increase the mite population. Spiralling whitefly was maximum during summer, dry weather conditions. Excess soil moisture due to high rainfall/ excess irrigation coupled with poor drainage and heavy soil conditions increases tuber rotting disease. Sweet potato whitefly and weevil are major pests affecting tuber yield and quality in sweet potato. Dry soil conditions promotes, whereas wet soil inhibits weevil infestation. Therefore, weevil population and tuber damage is minimum during *kharif* (rainy) season than during the summer season. However, effect of temperature and other climatic factors on the incidence and severity of pests and diseases of cassava and sweet potato is not well understood and needs to be studied in detail.

Information on response of cassava and sweet potato to high CO_2 and temperature is limited to few studies conducted abroad. In India, the impact of climate change (high CO_2, high temperature, UV-B) on the productivity and quality of cassava and sweet potato needs to be studied urgently.

References

Amthor, J.S. 2003. Efficiency of lignin biosynthesis: a quantitative analysis. *Annals. Bot.* **91**:673-95.

Angelov, M.N., Sun, J., Byrd, G.T., Brown, R.H. and Black, C.C. 1993. Novel characteristics of cassava, *Manihot esculenta* Crantz, a reputed C_3-C_4 intermediate photosynthesis species. *Photosynthesis Res.* **38**:61-72.

Balagopalan, C. 2000. Integrated technologies for value addition and PH management in Tropical tuber crops. pp. 147.

Bhattacharya, N.C., Bhattacharya, S., Strain, B.R. and Biswas, P.K. 1989. Biochemical changes in carbohydrates and proteins of sweet potato plants. *J. Plant Physiol.* **135**:61-66.

Bhattacharya, S., Bhattacharya, N.C. and Tolbert, M.E.M. 1990a. Characterization of carotene in sweet potato. (*Ipomoea batatas*) grown at CO_2 enriched atmosphere under field conditions. *In: Proceeding of Plant Growth Regulat. Soc. Am.*17th Annu. Meet. Plant. Growth Reg. Soc. Ithaca, NY R.H.,Hodgson (Ed.), pp.126-23.

Bhattacharya, N.C., Hileman, D.R., Ghosh, P.P., Musser, R.L., Bhattacharya, S. and Biswas, P.K. 1990b. Interaction of enriched CO_2 and water stress on the physiology and biomass production in sweet potatos grown in open top chambers. *Plant. Cell Environ* **13**:933-40.

Bhattacharya, N.C., Ghosh, P.P., Hileman, D.R., M.G. Huluka Alemayehu and Biswas, P.K. 1992. Growth and yield of sweet potato under different carbondioxide concentrations. *In: Sweet Potato Technology for the 21st Century*, Tuskegee Univ., Hill, W.A. Bonsi, C.K. and Loretan, P.A (Eds.), Tuskegee, AL. pp. 333-36.

Biswas, P.K., Hileman, D.R., Bhattacharya, N.C., Ghosh, P.P., Bhattacharya, S., Johnson, J.H. and Mbikayi, N.T. 1986. Response of vegetation to carbondioxide: Growth, yield and plant water relationships in sweet potatos in response to carbondioxide enrichment. Rep. 30. U.S. Dept. Energy. Carbondioxide Res. Div., Office of Energy Res., Washington DC.

Cen, Y.P. and Sage, R.F. 2005. The regulation of Rubisco activity in response to variation in temperature and atmospheric CO_2 partial pressure in sweet potato. *Plant Physiol.* **139:**979-90.

Clatayud, P.A., Baron, C.H., Velasquez, H., Arroyave, J.A. and Lamaza, T. 2002. Wild *Manihot* species do not possess C_4 photosynthesis. *Annals Bot.* **89:**125-27.

Edison, S. 1979. Cassava mosaic. *Annual Report*, CTCRI, Trivandrum, pp. 87-91.

Edwards, G.E., Sheta, E., D. Moore, B., Dai, Z., Franceschi, V.R., Cheng, S.H., Lin, C.H. and Ku, M.S.B. 1990. Photosynthetic characteristics of cassava (*Manihot esculenta* Crantz), a C_3 species with chlorenchymatous bundle sheath cells. *Plant Cell Physiol.* **31:**1199-1206.

El-Sharkawy, M.A. 2006. International research on cassava photosynthesis, productivity, eco-physiology and responses to environmental stresses in the tropics. *Photosynthetica* 481-12.

El-Sharkawy, M.A. and Cock, J.H. 1990. Photosynthesis of cassava (*Manihot esculenta*). *Exp. Agr.* 325-40.

El-Sharkawy MA, De Tafur SM and Cadavid LF. 1992. Potential photosynthesis of cassava as affected by growth conditions. *Crop Sci.* **32:**1336-42.

Harding, S.A., Guikema, J.A. and Paulsen, G.M. 1990. Photosynthetic decline from high temperature stress during maturation of wheat. *Plant Physiol.* **92:**648-53.

Imai, K., Coleman, D.F. and Yanagisawa, T. 1984. Elevated atmospheric partial pressure of carbondioxide and dry matter production of cassava (*Manihot esculenta* Crantz.). *Japan. J. Crop Sci.* **53:**479-85.

Irikura, Y., Cock, J.H. and Kawano, K. 1979. The physiological basis of genotype temperature interactions in cassava. *Field Crops Res.* **2:**227-39.

Mahon, J.D., Lowe, S.B., Hunt, L.A. and Thiagaraja, M. 1977. Environmental effects on photosynthesis and transpiration in attached leaves of cassava (*Manihot esculenta* Crantz). *Photosynthetica,* **11:**21-130.

Nawale, R.N. and Salvi, M.J. 1983. Effects of season on yield of sweet potato. *J. Root Crops* **9:**55-58.

Palaniswami, M.S., Mohandas, N. and Saraswathy, P. 1990. On farm survey on the incidence, extent and intensity of *Cylas formicarius* F. damage on sweet potato in Kerala. *J. Root Crops* **17:**180-96.

Palaniswami, M.S., Pillai, K.S., Nair, R.R. and Mohandas, C. 1995. A new cassava pest in India, *Cassava News Letter* **19:**6-7.

Palaniswami, M.S., Nair, R.R., Pillai, K.S. and Thankappan, M. 1996. Whiteflies on cassava and its role as vector of cassava mosaic disease in India. *J. Root Crops* **22:**1-8.

Pillai, K.S. and Daniel, R.S. 1979. Monthly variation in the population of whitefly (*Bemisia tabaci*) the vector of cassava mosaic disease. *J. Root Crops* **5:**8-10.

Pillai, K.S. and Palaniswami, M.S. 1985. *Spider mites of cassava.* Tech. Bul. 1, CTCRI, Trivandrum, pp. 17.

Pillai K.S. and Palaniswami M.S. 1990. Economic threshold and injury levels of spidermite damage on cassava. *J. Root Crops* **17:**220-22.

Rajamma P and Padmaja G.1981. Pests of sweet potato varietal resistance, biomass and control. Annual Report, CTCRI, Trivandrum, pp. 109-15.

Rajamma, P. and Pillai, K.S. 1987. Report on ecology and control of sweet potato weevil. IXth workshop on AICPTC, Assam Agri. Univ., Jorhat, 8-10 April.

Ramanujam, T. 1990. Effect of moisture stress on photosynthesis and productivity of cassava. *Photosynthetica* **24:**217-24.

Ravi, V. and James George. 2003. Studies on drought management in cassava. In: Annual Report, Central Tuber Crops Research Institute, Trivandrum. pp. 41-42.

Ravi, V. 2000. Photosynthesis and productivity of cassava under water deficit stress and stress free conditions. Session: II. Soil and Crop Management. In: *Int. Symp. Tropical Root and Tuber Crops*, 19-22 Jan., 200, CTRI, Trivandrum, Abst. No. 43, pp. 05.

Santhosh Mithra, V.S. and Somasundaram, K. 2008. A model to simulate sweet potato growth. *World Applied Sci. J.* (in press).

Ueno, O. and Agarie, S. 1997. The intercellular distribution of glycine decarboxylase in leaves of cassava in relation to the photosynthetic mode and leaf anatomy. *J. J. Crop Sci.* **66:**268-78.

Wrigley, C.W., Blumenthal, C., Gras, P.W., Barlow, E.W.R. 1994. Temperature variation during grain filling and changes in wheat-grain quality. *Aust. J. Plant Physiol.* **21:**875-85.

15

Impact of Climate Change on Vegetable Crops

N. K. Srinivasa Rao, R. H. Laxman and R. M. Bhatt

INDIA with diverse agro-ecological regions provides ample opportunity to grow a variety of horticulture crops round the year. Horticulture accounts for 28% of the agriculture GDP. Horticulture plays a pivotal role in sustaining small holdings, increasing employment opportunity, promoting export and above all achieving nutritional security. India is the world's second biggest producer of vegetables after China with 11.9% share in the total world production. India produced 109.05 million tonnes vegetables from an area of 7.16 million ha during 2005-06. In recent years, the rising incomes and greater health awareness, especially among urban Indians, have boosted the demand for vegetables, which needs to be met in a sustainable manner in future. Succulent vegetable crops are highly sensitive to climatic conditions.

Global warming and climate change is now unmistakably evident, from observations of average air and ocean temperatures, widespread melting of polar snow and ice and a rise in the average sea-level worldwide. According to the Intergovernmental Panel on Climate Change (IPCC) Report 2007, the global mean surface temperatures have increased by 0.74 °C ± 0.18 °C during the last 100 years (1906–2005). The rate of warming during the last 50 years is almost double that for the last 100 years. The earth temperature may increase by 1.1 to 6.4 °C during 2090–99 compared to 1980–99 in different regions of the world. The report also indicates that crop yield losses would be severe in tropical regions. Many regions are already experiencing, which may worsen as mean temperatures rise and climate becomes more variable and unpredictable. Precipitation intensity and patterns would also change leading to both droughts and floods. The warmer climate would increase evaporative demand of crops. Another dimension of climate change is the elevated CO_2 in the atmosphere and physiological response of different crops. The interaction of elevated CO_2 and warming would determine the extent of impact of climate change on horticulture sector in general and vegetable production in particular over an agro-ecological region.

Climate change presents both adverse and beneficial impacts on crop production depending upon the agro-ecological region, crop and the season. Understanding the likely impacts of global warming and climate change on crop growth, development and in turn productivity of vegetable crops is a first step in order to prepare ourselves in containing the impact of climate change on horticulture sector.

IMPACT ANALYSIS ON VEGETABLE CROPS

Effect of Elevated CO_2

Increased concentration of atmospheric carbon dioxide stimulates crop growth by the so-called "Carbon fertilization" effect (Rogers and Dahlman, 1993). A doubling of the atmospheric CO_2 concentration increases plant growth and marketable yield by 30–40% in C3 plants (Kimball, 1983; Poorter, 1993). Most plants growing in enhanced CO_2 exhibit increased rates of net photosynthesis brought about by increased availability of CO_2 at the chloroplasts and reduced photorespiration resulting from an increased ratio of O_2 to CO_2 (Farquhar and Sharkey, 1982, Pearcy *et al.*, 1987). The higher photosynthesis rates are then manifested in higher leaf area, dry matter production and yield for many crops (Kimball, 1983; Acock and Allen, 1985).

Productivity

In cucumber an yield increase of 34% was observed at 620μ mol mol^{-1} CO_2 compared with a crop grown at 364 μ mol mol^{-1} CO_2 (Nederhoff, 1994). In bell pepper increasing CO_2 from 300 μ mol mol^{-1} CO_2 to 450 μ mol mol^{-1} CO_2 increased fruit production by 46%. One/ third of this increase was caused by a shift in allocation, and the remaining two thirds was caused by increased CO_2 assimilation rate.

In tomato and eggplant, elevated CO_2 increased yield by 31 and 24%, respectively. However, disorders of leaf and shoot growth occurred and active leaf area was reduced. In two onion cultivars increase in CO_2 concentration shortened time to bulbing but prolonged the period from bulbing to bulb maturity with bulb yield increase of 28.9–51.0% due to increase in the rate of leaf area expansion and the rate of photosynthesis during the pre-bulbing (Daymond *et al.*, 1997). Elevated CO_2 (550 ppm) influenced overall growth and development of onion (cv. Arka Kalyan) producing higher dry matter content in leaves, stems and bulbs at IIHR, Bangalore (Table 15.1). Enhanced accumulation of dry matter was translated into bigger-sized bulbs with an average diameter of 75.02 mm under elevated CO_2 compared to 65.67 mm under ambient levels. The average bulb weight was 171.98 g and 117.96 g, in elevated and ambient CO_2 levels, respectively. The yield increased by 25.9 % at elevated CO_2 over the control mainly due to increase in bulb size. At bulb development stage, photosynthesis rate was 17.74 μ mol m^{-2} s^{-1} at elevated levels as compared to 9.48 μ mol m^{-2} s^{-1} under ambient levels. The transpiration rate was lower (15.30 m mol m^{-2} s^{-1}) as compared to 19.04 m mol m^{-2} s^{-1} and stomatal conductance was also lower (2.71 mol m^{-2} s^{-1}) as compared to 4.02 mol m^{-2} s^{-1} at elevated and ambient levels of CO_2, respectively.

In tomato cv. Arka Ashish elevated CO_2 (550 ppm) influenced growth and development and increased yield by 24.4% (Table 15.2). Among growth parameters plant height, number of secondary branches and leaf area were higher at elevated level as compared to ambient levels at fruiting stage. The dry matter accumulation in leaf, stem and fruits was higher at elevated CO_2 level as compared to ambient level. The photosynthesis rate was higher 29.29 μ mol m^{-2} s^{-1} at elevated CO_2 compared to 19.12 μ mol m^{-2} s^{-1}, but there was a reduction in transpiration rate and stomatal conductance. The fruit yield/plant was also higher (3.46 kg/plant) at 550 ppm as compared to 2.75 kg/plant under ambient conditions. The yield increase was mainly due to increase in the number of fruits/plant at elevated CO_2 (74 fruits/plant) compared to chamber control (56 fruits/plant).

Table 15.1. Effect of elevated carbon dioxide on growth and gas exchange characteristics of onion Arka Kalyan

	Bulb initiation stage		*Bulb development stage*	
	550 ppm	*Ambient*	*550 ppm*	*Ambient*
Pseudo stem height (cm)	4.6	3.76	13.8	11.0
Number of leaves	0.5	4.2	9.8	9.4
Dry weight of leaf (g)	0.31	0.18	4.58	3.22
Dry weight of stem (g)	0.18	0.05	1.63	0.91
Dry weight of bulb (g)	0.07	0.05	7.84	4.06
Leaf area (cm^2)	100.31	60.02	775.92	564.88
Photosynthesis rate (mol m^{-2} s^{-1})	11.81	9.89	17.78	9.48
Transpiration rate (mm m^{-2} s^{-1})	2.82	3.31	15.30	19.04
Stomatal conductance (mol m^{-2} s^{-1})	0.16	0.22	2.71	4.02

Fig. 15.1. Onion crop inside the Open Top Chamber at 550 ppm CO_2

Table 15.2. Effect of elevated carbon dioxide on growth and gas exchange parameters of tomato cv. Arka Ashish at fruiting stage.

	550 ppm	Ambient (350 ppm)
Plant height (cm)	114	102.5
Number of primary branches	11	12
Number of secondary branches	65.5	47.5
Dry weight of leaf (g)	222.10	110.02
Dry weight of stem (g)	155.38	78.53
Dry weight of fruits (g)	149.26	137.64
Leaf area (m^2)	1.31	1.28
Photosynthesis rate(mol m^{-2} s^{-1})	27.44	18.15
Transpiration rate(mm m^{-2} s^{-1})	11.87	13.01
Stomatal conductance (mol m^{-2} s^{-1})	0.34	0.48

In most crops elevated increased CO_2 improves WUE due to reduced stomatal conductance, potentially decreasing drought susceptibility and reducing irrigation requirements. However, effect of decreased transpiration on vegetable crop yields is unlikely to be large since vegetables are mostly irrigated. Physiological disorders, blossom end rot in tomato, pepper and watermelon are sometimes associated with excessive transpiration, and so the incidence of these disorders may be reduced.

Fig. 15.2. Tomato crop inside the Open Top Chamber

Effect of rise in Temperature

The positive effect of elevated CO_2 might be offset by the adverse effect of associated global warming particularly excessive heat and drought. Almost all horticultural plants are highly sensitive to the direct effects of rise in temperature and precipitation. In crops like tomato and potato, CO_2 enrichment does not appear to compensate for the detrimental effects of higher temperatures on yield. In general, crops with a higher harvest Index, high sink demand, indeterminate growth and long growth seasons are considered most likely to respond to the combination of higher CO_2 and temperature.

Temperature affects horticultural crops in several ways, by influencing crop duration, flowering, fruit growth, ripening and quality. The impact of rising temperature is likely to differ with region and type of the crop. The temperature rise may not be evenly distributed between day and night and between summer and winter. Most models, theories and observations suggest that night minimum will increase more than daytime maximum and winter temperatures will increase more than summer temperatures. In tropical regions, even moderate warming may lead to disproportionate declines in yield. In higher latitudes, however, crop yields may improve as a result of a small increase in temperatures. Temperatures in developing countries, which are predominantly located in lower latitudes, are already closer to or beyond thresholds and further warming will reduce rather than increase productivity. For most vegetables growth is more rapid as temperatures increase, at least up to 25 °C. High temperature injuries commonly reduce productivity.

A greater increase in winter temperatures than summer temperatures should reduce the potential for summer heat stress, but may lead to a lack of vernalization for some crops. Thus, if winters become milder, different planting dates and cultivars may be required. Vegetables are generally sensitive to environmental extremes, and thus high temperatures and limited soil moisture are the major causes of low yields in the tropics and warming will influence the severity of environmental stress imposed on vegetable crops. The sensitivity of individual crops to temperature depends on the inherent tolerance and growing habits. Hall and Allen (1993) suggested that indeterminate crops are less sensitive to periods of heat stress because the time of flowering is extended compared with determinate crops.

Vegetative and reproductive processes in tomato are strongly modified by temperature alone or in conjunction with other environmental factors (Abdalla and Verkerk, 1968). High temperatures affect reproductive development in two ways, both of which potentially reduce yields. Firstly the rate of reproductive development is accelerated, which shortens the fruit maturation period, resulting in lower fruit weight. In many crops, reproductive events themselves are prevented at temperatures only a few degrees above optimal. In tomato, high temperatures can cause significant losses in productivity due to reduced fruit set, with smaller size and lower quality fruits (Stevens and Rudich, 1978). Temperature influences the dry matter partitioning between fruits and vegetative plant parts either directly or indirectly through its influence on development, flower and or fruit abortion. Hazra *et al.,* (2007) summarized the symptoms causing fruit set failure at high temperatures in tomato; this includes bud drop, abnormal flower development, poor pollen production, dehiscence, and viability, ovule abortion and poor viability, reduced carbohydrate availability, and other reproductive abnormalities. The optimal temperature for net assimilation rate in tomato is between 25–30 °C and optimal daily mean temperatures for fruit set of tomato have been reported as 21–24 °C (Geisenberg and Stewart, 1986). In tomato, high temperatures after pollen release decreased fruit setting, yields and seed set even when pollen was produced under optimal conditions (Peet *et.al.*, 1997). High temperature above 28 °C induced maximum flower and fruit drop in tomato. Cell Membrane stability, measured as percentage injury to leaf tissues grown at high temperature conditions indicated that CV. CL 5915 – 223 – 2 1 – 01 with minimum injury had about 65% fruit set. High temperature tolerant tomato line from AVRDC CL – 5915 – 206DG 2 – 2 – 0 had significantly higher leaf area and total biomass than susceptible CV. Arka Saurabh grown at 35 °C day and 27 °C night temperature (Srinivasa Rao, 1995). Reduced fruit setting in tomato (Kinet and Peet, 1997) and pepper (Wien, 1997a) occurs as a result of high temperature. Overall pre-anthesis stress appears to be more injurious than stress applied after pollen arrives on the stigma.

In pepper, high temperature exposure at the pre-anthesis stage did not affect pistil or stamen viability, but high post-pollination temperatures inhibited fruit setting, suggesting that fertilization is sensitive to high temperature stress (Erickson and Markhart, 2002).

Bulbing in onion is induced by photoperiod, once induced it occurs rapidly at higher temperatures (Brewster, 1997). In pea temperatures above 25.6 °C during bloom and pod set reduce flower and pod number and yields. In beans high temperatures delays flowering because they enhance the short day photoperiod (Davis, 1997). Increasing temperature generally increases the developmental rate, but the growth rate is not necessarily stimulated to the same degree (Krug, 1997). In an early maturing pea cultivar, increasing the temperature from 16–24 °C increased the differentiation rate of nodes from emergence to the appearance of the first flower, but the growth rate was reduced, which reduced total dry matter and yields.

In cucumber sex expression is affected with low temperatures leading to generally desirable more female flowers and high temperatures leading to production of more male flowers (Wien, 1997b). In onion, warmer temperatures shortened the duration of growth, accounting for a negative correlation between elevated temperatures and crop yields (Daymond *et al.*, 1997).

INTERACTION EFFECT OF ELEVATED CO_2 AND TEMPERATURE

Temperature increases due to global warming will occur simultaneously with increase in CO_2 and their interaction would determine the extent of impact on vegetable production. Stomatal closure from high CO_2 may reduce water losses. On the other hand transpiration has also a cooling effect on leaf temperature. If transpirational cooling is reduced, leaves could become more vulnerable to heat stress. In general crops with a higher Harvest Index, high sink demand, indeterminate growth and long growth seasons are considered most likely to respond to the combination of higher CO_2 and temperature.

In onion an increase in bulb yield due to a rise in CO_2 from 374 to 532 μ mol mol^{-1} was offset by a temperature warming of 8–10 °C in one cultivar and 4–5.8 °C in another cultivar. Differences in cultivar responses may be due to the temperature sensitive cultivar being a short season type. In a temperature gradient tunnel study on Sito and Hysam, the rates of progress from transplanting to bulbing, and from bulbing to harvest maturity, were positive linear functions of mean temperature. At a given temperature, the time of bulbing was earlier, but the duration of bulb growth longer, at elevated compared with normal CO_2. Radiation-use efficiency was greater at elevated than at normal CO_2 concentrations in the period up to bulbing, but was the same at both CO_2 concentrations during subsequent bulb growth. Total crop dry weight at bulbing increased by 32–44% due to elevated CO_2 concentration. Bulb yields at harvest maturity declined with progressively warmer temperatures and to a greater extent in Sito than Hysam. Enrichment with CO_2 increased bulb yields by 29–37% and by 35–51% in Hysam and Sito, respectively. From comparison of the temperature rise needed to offset entirely the yield increases in each cultivar due to elevated CO_2, it is concluded that current estimates of climate change should be beneficial for bulb onion production, particularly for long season cultivars (Daymond *et al.*, 1997).

The possible impacts of rise in atmospheric CO_2 concentration and temperature on beetroot, carrot and bulb onion plants with CO_2 concentrations of 350, 450, 550, 650 and 750 vpm and temperatures of 12, 13.5, 15, 16.5 and 18 °C were examined. Measurements of the ratio of maximum diameter of bulb to minimum diameter of neck for onions showed that there was little or no influence of CO_2, whereas the effect of temperature was substantial. Bulbing was accelerated by high temperature and was greatly delayed at low temperature. At temperatures <15 °C, the delays to bulbing resulted in development of undesirable, thick-necked onions which tended to remain green with erect leaves. The results suggested that a warmer climate will be advantageous for commercial production of bulb onions in Britain (Wurr, 1998).

Quality

The data on quality parameters of onion showed a decrease in antioxidant content at elevated CO_2. Total phenols were higher and a decrease in flavonoids content was observed. Quality parameters of tomato showed significant increase in lycopene and carotenoid content at elevated CO_2. Reduced sugar content in peas is often attributed to increased night time respiration, although it may also be caused by shorter period over which the fruit develops at high temperatures.

Pests of Target Crop

Information on climate change and global warming on pests of vegetable crops is severely lacking.

EFFECT OF DROUGHT AND FLOODING ON VEGETABLE CROPS

Indian climate is dominated by the monsoon responsible for most of the region's precipitation. It is critical for the availability of drinking water and irrigation for agriculture. The variability in rainfall frequency and distribution patterns associated with climate change would bring about changes in soil moisture availability, consequently crop water requirement.

Vegetables, being succulent products by definition, generally consist of greater than 90% water (AVRDC, 1990). Thus, water greatly influences the yield and quality of vegetables; drought and flooding drastically reduce vegetable productivity. Drought stress causes an increase in solute concentration of plant cells, reducing the water potential and disruption of membranes and several processes occurring in the cell. Shallow rooted vegetables are known to be sensitive to water deficiency and require frequent supply of moisture for better yield and quality.

In onion, soil-water stress caused by withholding irrigation at both the 3 and 7 leaf stages reduced yields by 26% compared to control.

In beans water stress lower than -40 hPa decreased leaf water potential and stomatal conductance and it also reduced all growth and yield parameters. This effect was significant, especially at -150 and -200 hPa. Plants at -40 hPa grew normally while plants at -100 hPa showed some drought acclimation (El-Tohamy, 1999).

Response of 20 selected cultivars of tomato to imposed moisture stress at vegetative and fruiting stage and complete stress after establishment indicated that in completely stressed plants, about 20–50% fruit drop was observed in different cultivars. Water stress accompanied by high temperatures above 28 °C during fruiting induced about 30–45% flower drop in different cultivars. The cultivars, PED, Redrock, KT-1, Arka Saurabh and Sioux, had more negative water and osmotic potential and recovery capacity was better in these cultivars (Srinivasa Rao and Bhatt, 1992). The differential growth of root and shoot of tomato was related to osmotic adjustments (Shashidhar *et al.*, 1991). Tomato, Arka Meghali has better osmotic adjustment under rainfed conditions and crop water stress index (CWSI) was low in Arka Meghali (0.76) at flowering and fruit development stages compared to other cultivars (0.80–0.96) (Anonymous,1996). A reduction of 17–26% in yield was observed under moisture stress in Arka Niketan and N-2-4-1, while minimum yield reduction (15%) was found in SM-11. The study indicated that genotype SM-11 that has smaller bulb size and narrow and upright leaf orientation performed better than other two genotypes, which have relatively big bulb size, and large leaves under different levels of moisture stress (Bhatt *et al.*, 2006).

Vegetable production in our country is taken up both during rainy and dry seasons. During rainy seasons due to excess rains, the flooding occurs in vegetable fields and vegetables are generally unable to tolerate excessive soil moisture. Most vegetables are highly sensitive to flooding and damage is due to reduction of oxygen in root zone which inhibits aerobic processes. Flooded tomato plants accumulate endogenous ethylene that causes damage to the plants (Drew, 1979). Low oxygen levels stimulate an increased production of an ethylene precursor, 1-aminocyclopropane-1-carboxylic acid (ACC), in roots. The rapid epinastic growth of leaves is a characteristic response of tomatoes to waterlogged conditions and the role of ethylene accumulation has been implicated (Kawase, 1981). The severity of flooding symptoms increases with rising temperatures; rapid wilting and death of tomato plants is usually observed following a short period of flooding at high temperatures (Kuo *et al.,* 1982).

Aberrant Weather and Extreme Events

Observations on aberrant weather and extreme events and their effect on vegetable crops are not well documented and it needs to be done for possible adaptation measures in future.

Adaptation Measures for Climate Change and Global Warming

- Cultural management like changing sowing or planting dates and modified manuring and fertilization for conservation of soil organic matter.
- Fertigation, drip and sprinkler irrigation and conservation of soil moisture through mulching with crop residues and plastic mulches.
- Growing vegetables in raised beds to avoid flooding.

Mitigation Measures to Reduce Emission of CO_2/GHG and Carbon Sequestration Potential of Crops

Most of the vegetables being annual crops do not have any carbon sequestration potential, the scope for reducing emissions in their cultivation is highly limited, and information on these aspects is lacking. Other mitigation measures are as follows:

- Resource conservation techniques
- Organic farming.

FUTURE STRATEGIES FOR RESEARCH

- Breeding vegetable hybrids and varieties tolerant to heat and drought stress.
- Compilation and application of indigenous technical knowledge of vegetable production under warm and drought conditions.
- Develop agro-techniques for warm weather cultivation.
- Development and validation of vegetable crop model for Indian conditions.
- Studies on impact assessment of climate change and global warming on vegetable crops.
- Studies on impact of climate and global warming on dynamics of pest complex of vegetable crops.

CONCLUSION

Recently the demand for vegetables in India has been growing due to rising incomes and greater health awareness, which needs to be met in a sustainable manner in future. The succulent vegetable crops are highly sensitive to climatic conditions, heat, drought and flooding. Direct and indirect observations indicate increased levels of atmospheric carbon dioxide and rise in temperature from 1–5 °C by the end of this century with changing rainfall pattern and greater frequency of extreme events of drought and floods. Elevated CO_2 has positive effect ranging from 24–51% on productivity of vegetable crops like cucumber, bell pepper, tomato, eggplant, and onion. However, rise in temperature affects crop duration, flowering, fruiting, fruit size, and ripening of vegetable crops with reduced productivity and economic yield. Therefore, overall impact of climate change and global warming will depend on interaction effect of elevated CO_2 and temperature rise. In most vegetable crops, the interaction effect does not appear to be positive except onion. Accurate impact analyses of global warming on vegetable crops are required to evolve adaptation measures and future strategies to cope with climate change and global warming.

References

Abdalla, A.A. and Verderk, K. 1968. Growth, flowering and fruit set of tomato at high temperature. *The Netherland J. Agric Sci.* **16**:71-76.

Acock, B. and Allen, Jr. A.H. 1985. Crop responses to elevated carbon dioxide concentrations. In: B.R. Strain and J.D. Cure (eds.), "*Direct Effects of Increasing Carbon Dioxide on Vegetation DOE/ER-0238*", U.S Department of Energy, Washington, D.C, pp.53-97.

AVRDC. 1990. Vegetable Production Training Manual. Asian Vegetable Research and Training Center. Shanhua, Tainan, 447 pp.

Bhatt, R.M., Srinivasa Rao, N.K. and Veere Gowda, R. 2006. Response of bulb onion (*Allium cepa* L.) to water stress: photosynthesis, stomatal conductance and osmotic adjustment. *Indian J. Horticulture* **63:**276-80.

Brewster, J.L.1997. Onion and Garlic. *In: The Physiology of Vegetable Crops.* Wien, H.C. (Ed.) CAB International, UK, pp.581-20.

Davis, J.H.C. 1997. Phaseolous beans. *In:The Physiology of Vegetable Crops.* CAB International, UK, pp .409-28.

Daymond, A.J., Wheeler, T.R., Hadley, P., Ellis, R.H. and Morison, J.I.L. 1997. The growth, development and yield of onion (*Allium Cepa* L.) in response to temperature and CO_2. *J. Hort. Sci.***72:**135-45.

Drew, M.C. 1979. Plant responses to anaerobic conditions in soil and solution culture. *Curr Adv Plant Sci.* **36:**1-14.

Erickson, A.N. and Markhart, A.H. 2002. Flower developmental stage and organ sensitivity of bell pepper (*Capsicum annuum* L.) to elevated temperature. *Plant Cell Environ.* **25:**123-30.

Farquhar, G.D. and Sharkey, T.D. 1982. Stomatal Conductance and Photosynthesis, *Annu. Rev. Plant Physiol.* **33:**317-45.

Geisenberg, C. and Stewart, K. 1986. Field Crop management. *In: The Tomato Crop, A Scientific Basis for Improvement.* Atherton, J.G. and Rudich, J. (Eds.). Chapman and Hall, New York, pp.1217-29.

Hall, A.E. and Allen, L.H. Jr. 1993. Designing cultivars for the climatic conditions of the next century. *In: International Crop Science I.* Buxton, D.R., Shibles, R. Forsberg, R.A., Blad, B.L., Assay, K.H., Paulsen, G.M. and Wilson, R.F. (eds). Crop Science Society of America. Madison, Wisconsin, pp. 291-97.

Hazra, P., Samsul, H.A., Sikder, D. and Peter, K.V. 2007. Breeding tomato (*Lycopersicon Esculentum* Mill) resistant to high temperature stress. *International Journal of Plant Breeding* **1**(1).

Idso, S.B., Kimball, B.A., Anderson, M.G. and Mauney, J.R. 1987. Effects of atmospheric CO_2 enrichment on plant growth: the interactive role of air temperature. *Agril. Ecosys. Environ.* **20:**1-10.

Kawase, M. 1981. Anatomical and morphological adaptation of plants to waterlogging. *Hort. Science* **16:**30-34.

Kimball, B.A. 1983. Carbon dioxide and agricultural yield: an assemblage and analysis of 430 prior observations. *Agron. J.* **75:**779-88.

Kinet, J.M. and Peet, M.M. 1997. Tomato. In: *The Physiology of Vegetable Crops.* C. Wien(Ed.). CAB International, New York, NY. pp. 207-58.

Krug, H. 1997. Environmental influences on development, growth and yield. *In: The Physiology of Vegetable Crops.* Wein, H.C. (Ed.) CAB International, Wallingford, UK, pp. 101-80.

Kuo, D.G, Tsay JS, Chen BW and Lin P.Y. 1982. Screening for flooding tolerance in the genus. *Lycopersicon. Hort. Science* **17:**76-78.

Nederhoff, E.M. 1994. Effects of CO_2 concentration on photosynthesis, transpiration and production of greenhouse fruit vegetable cops. Ph.D. Dissertation, Wageningen, the Netherlands, pp. 213.

Pearcy, R.W., Bjorkman, O., Caldwell, M.M., Keely, J.E., Monson, R.K. and Strain, B.R. 1987. Carbon gain by plants in natural environments, Bio *Science* **37:**21-29.

Peet, M.M., Willits, D.H. and Gardner, R. 1997. Response of ovule development and post-pollen production process in male sterile tomatoes. *Plant Cell and Environ.* **21:**225-31.

Poorter, H. 1993. Interspecific variation in the growth response of plants to an elevated ambient CO_2 concentration. *Vegetatio* **104/105:**77-97.

Rogers, H.H. and Dahlman, R.C. 1993. Crop responses to CO_2 enrichment. *Vegetatio* **104/105:**117-31.

Shashidhar, V.R., Mervat, E.S., Hebbar, B.K., Prasad, T.G. and Udayakumar, M. 1991. Root osmotic adjustment promotes higher water status in shoots. Evidence for a link between drought tolerance and avoidance mechanism in two crop species. "*National Seminar of Young Scientists on Environmental Physiology*", Garhwal, Srinagar, pp. 22.

Srinivasa Rao, N.K. and Bhatt, R.M. 1992. Responses of tomato to moisture stress : Plant water balance and yield. *Plant Physiol. & Biochem.* **19:**36-41.

Srinivasa Rao, N.K. 1995. Management of heat moisture and other physical stress factors in tomato and chilli in India. In collaborative vegetable research in South Asia, "*Proceedings of the SAVERNET Mid term Review Workshop*". AVRDC, Shanhra, Tainan, Taiwan, pp. 232.

Stevens, M.A. and Rudich, J. 1978. Genetic potential for overcoming physiological limitations on adaptability, yield, and quality in tomato. *Hort. Science* **13:**673-78.

Weis, E. and Berry, J.A. 1988. Plants and high temperature stress. *Soc. of Expt Biol*, pp 329-46.

Wien HC. 1997a. Pepper, *In: The Physiology of Vegetable Crops*. Wien, H.C. (Ed). CAB International, Wallingford, UK, pp.259-93.

Wien, H.C. 1997b. The Cucurbits: Cucumber, melon, squash and pumpkin. *In: The Physiology of Vegetable Crops*. Wien, H.C. (Ed). CAB International, Wallingford, UK, pp. 345-86.

Wurr, D.C.E., Hand, D.W., Edmondson, R.N., Fellows, J.R., Hannah, M.A. and Cribb, D.M. 1998. Climate change: a response surface study of the effects of CO_2 and temperature on the growth of beetroot, carrots and onions. *J. Agric Sci.* **131:**125-33.

16

Impact of Global Warming on Mushroom Cultivation and Integrated Pest and Disease Management Strategies

V. P. Sharma, R. P. Tewari and S. R. Sharma

GLOBAL Warming is defined as the increase of average temperature on earth. The change in global climate is brought by the enhanced atmospheric carbon dioxide (CO_2), tropo-and stratospheric ozone (O_3) levels, ultraviolet radiation *etc*. In addition, agricultural and industrial activities also add considerable amounts of gases like methane (CH_4), nitrous oxide (N_2O) and chlorofluorocarbons (CFCs) to the atmosphere. Collectively, all these gases lead to greenhouse effect, because all of them absorb infrared radiation. It is estimated that about 50% of the CO_2 emitted to atmosphere remains in it and other 50% is taken up by the ocean and terrestrial ecosystem (Houghton *et al.*, 1990). As the earth is getting hotter, extreme events like hurricanes, droughts and floods are getting more frequent thereby, affecting, agricultural production and supply pattern in India. Over the last 100 years, the average temperature of air near the earth's surface has risen a little less than 1 °C (0.74 ± 0.18 °C). Out of the 20 warmest years on record, 19 have occurred since 1980. The three hottest years ever observed have all occurred in the last eight years. Effects of climate change on ecosystem, fauna and flora are also noticeable. It will affect the crop cycles and incidence of pests and diseases will increase. Adaptation to climate change requires reframing of planning and policy according to the new emerging situation.

Mushrooms play an important role in the environment. They can be grown on many of the so called "agricultural wastes" — rice straw, saw dust, banana pseudo-stems, peanut shells, corn cobs, vegetable trimmings, sugarcane bagasse, fruit pericarps, cotton waste and many other waste material of plant origin like paper and fabric scraps. These substrates are eventually broken down during the growth of mushroom and become compost which can be applied to field. Thus, mushroom production can be incorporated into an existing integrated agricultural or farming system.

The use of agricultural wastes as mushroom substrates will reduce the burning of these wastes and restrict environmental pollution. Cleaning fields, roadsides and forest margins of plant refuse and using these as mushroom substrates has the added advantage of preventing forest fires. Finally, an appreciation of edible cultivated mushrooms will increase awareness of wild relatives and hopefully lead to the protection of natural mushroom flora and their habitats. Increase in crop production is resulting in agricultural wastes in abundance throughout the world creating environmental pollution. India produces about 600 million tonnes of agro-wastes per annum which includes about 130 million tonnes of wheat straw, 85 million tonnes of paddy straw, 186 million tonnes of sugarcane bagasse, 30 million tonnes of maize straw, 36.24 million tonnes of millet straw, 13.14 million tonnes of pulses straw, 25.58 million tonnes of oilseeds waste, 16.19 million tonnes of cotton waste and rest from coconut, jute and other crops. Besides this good quantity of waste is available from agro-industries and poultry which can also be utilized for mushroom production.

IMPACT ANALYSIS ON MUSHROOM CULTIVATION

Elevated CO_2 and Temperature

Mushroom cultivation is an indoor activity and its cultivation is done in controlled environmental conditions. Therefore, most of the effects of climate change on mushroom are indirect and not the direct outcome of elevated CO_2 and rise in temperature.

Productivity

In India, mushroom cultivation is done both seasonal as well as under controlled environmental conditions. Mushroom cultivation undertaken as seasonal activity is likely to be affected by change in the climate. The indirect effect of climate change on mushroom production is listed below.

- Survey conducted in different mushroom farms in Haryana and Punjab revealed that growers are facing more problems of stroma formation and diseases in October and November in those years when temperature remains high. Therefore, it is limiting the production period of white button mushroom under natural conditions (seasonal).
- Under controlled environmental conditions the production cost is increasing.
- Growers are compelled to grow low value mushroom species.
- Mushroom biodiversity is facing extinction forever due to continuing decline in their habitat because of deforestation, population explosion and the global warming. Regular survey conducted revealed that occurrence of mushroom species under natural conditions is being affected. The effects are more disastrous due to forest fires. It is affecting the germplasm availability, breeding programmes and diversification.

Quality

Higher temperatures during October-November in plains results in abnormality like stoma formation, pin head death and browning of white button mushrooms cultivated under natural conditions (seasonal crop).

Pest of Target Crop

In white button mushrooms higher temperature (22–26 °C) during October–November in plains increases the incidence of various competitors like false truffle (*Diehliomyces microsporus),* olive green mould (*Chaetomium olivaceum),* brown plaster mould (*Papulaspora byssina),* yellow mould (*Myceliophthora lutea, Chrysosporium luteum, C. sulphureum*), ink caps (*Coprinus* spp.), green mould diseases (*T. viride, T. koningii, T. hamatum, T. harzianum,),* diseases like dry Bubble (*Verticillium fungicola* var. *fungicola*), wet bubble (*Hypomyces perniciosus),* cobweb (*Cladobotryum dendroides),* bacterial blotch (*Pseudomonas tolaasii),* pests like (sciarids (*Lycoriella agarici, L. mali, Bradysia agrestis, B.coprophila B. paupera),* phorids (*Megaselia halterata, M. nigra,* and *M. agarice. Megaselia sandhui),* cecid flies (*M. speyeri, M. fungicola, M. barnesi,* mites (*Tyroglyphus putrescentiae, T.dimidiatus, Histiostoma heinemanni, H. miles, Hypoaspis mali, Pygmephorus* sp., *Bakerdania* sp., *Parasitus fumetorus* and *P.consanguineus)* and nematodes (*Ditylenchus myceliophagous Aphelenchoides, agarici, A. composticola, A. myceliophagus, A. neocomposticola, A. sacchar).* Besides, abundance of ink caps on beds and mouldy appearance on cultivation substrate of paddy straw mushroom is also noticed due to high temperature. When the temperature goes high it takes very long time for the appearance of fruit bodies after the bags are opened and some times very low production/a few fruit bodies are formed in oyster mushroom. Obvious insect damage and high incidence of cob web (*Cladobotrym apiculatum, C. verticillatum), Gliocladium* spots (*Gliocladium virens, G.deliquescens*) and yellow blotch (*Pseudomonas agarici)* are also high due to high prevailing temperature during the cultivation of oyster mushroom.

Some diseases particularly false truffle, bacterial blotch are worth mentioning here. False truffle is main limiting factor in the cultivation of *Agaricus bitorquis* which is morphologically almost identical to the commonly cultivated white button mushroom. The disease is of common occurrence during February or early March in *A. bisporus* in plains of Northern India and during summer months in *A. bisporus* and *A. bitorquis* in hilly regions of the country. Ascospore production is abundant at 25 and 30 °C but not at 15 °C. Optimum growth of the fungus has been recorded at 26–28 °C. It is mainly a disease of *A. bitorquis* wherein crop is raised at 25 ± 1 °C but it also develops very fast in *A. bisporus* when crop is taken under natural climatic conditions and temperature rises above 20 °C. Cultivation technology of *Agaricus bitorquis* has been standardized about 15 years back but still no body is growing this mushroom due to the occurrence of false truffle. Therefore, work must be initiated to manage this disease either through IPM or developing resistant strains. Bacterial blotch infecting both button as well as oyster mushroom creates havoc as and when temperature goes high and fruit bodies remain wet for at least two hours.

Similarly, among pests especially phorids and sciarids are rising drastically and are hampering the mushroom production both ways directly by feeding on the crop and indirectly by transmitting various diseases.

Phorids: The optimum temperature for phorid flies is 23–25 °C. At 25 °C incubation, larval and pupal periods last for 1.6, 4.6 and 7.3 days, respectively. Developmental period from egg to adult is 14.6 days. Lower temperature increases the generation time. Seasonal abundance of phorid flies during different years in NRCM mushroom farm is presented in Table 16.1.

Table 16.1. Seasonal abundance of phorids during different year

Month	1999 Phorids		2000 Phorids		2001 Phorids	
	Temperature (°C)	Population	Temperature (°C)	Population	Temperature (°C)	Population
Jul	25.8	4310	25	334	25.2	938
Aug	23	1830	23.5	1423	23	1137
Sept	22.2	174	18.5	554	18	528
Oct	18	5	18.6	104	18	742
Nov	16	0	16.2	6	15	113
Dec	11.8	0	14.7	0	14.7	2
Jan	-	-	11.2	4	10	0
Feb	-	-	10.7	0	14	0
Mar	-	-	15.2	0	15.2	0
Apr	-	-	19	0	20	0
May	-	-	22	0	21.7	44
Jun	-	-	24	110	21.6	1816

Source: Kumar and Sharma (2001)

The results presented in Table 16.1 clearly showed that temperature range 23–25 °C is highly congenial for the development of phorid flies.

Sciarids: At 7.2, 12.8, 18.3 and 23.9 °C temperature mean number of eggs laid per female was observed to be 12.6, 48.4, 58.8 and 62.5, respectively.

In ***Pleurotus*** both low temperature and high temperature strains are available for cultivation. But mushroom flies are the major constraint for the cultivation of high temperature strains.

Aberrant Weather and Extreme Events

Observations on extreme events and aberrant weather on mushrooms are lacking, because the crop is mostly grown under controlled conditions and only suitable seasons. However, incidences of forest fires due to global warming have become frequent affecting the germplasm availability, breeding programmes and diversification. Mushroom biodiversity is facing extinction due to continuing decline in their habitat because of deforestation, population explosion and the global warming. The production period of some of the mushrooms under natural conditions is observed to be restricted to a limited period only.

Adaptation Measures for Climate Change and Global Warming

Preserving existence and distribution of fauna and flora: Overall existence and distribution of fauna and flora and some species are already facing extinction. Preserve forest areas, avoid forest fires and collection of some endangered species may be restricted or may not be permitted at all. Establish mycodendrons namely, *Tuber melanosporum, Boletus edulis, Cantharellus cibarius, Paxillus involutus, Scleroderma vulgare, Pisolithus tinctorius* and others. Establishment of mycodendrons will not only lead to the production of highly valued edible mushroom but also improve the forest health by giving the tree very substantial protection by diverse mechanisms like increased efficiency of photosynthesis, enabling tree to transpire if the ground is dry, increased nutrient and water absorbing area and ensured nitrogen supply to the tree.

Developing short duration varieties: Seasonal cultivation of various mushrooms will be affected. The production period for cultivation of button mushroom and other low temperature requiring species will be limited. Growers may be compelled to abandon the cultivation of these species. There is need to develop short duration varieties of *Agaricus bisporus* or modify the production technology to quick harvest through nutrient management. Introduce and commercialize *Agaricus bitorquis,* the high temperature tolerant sister species of *A. bisporus.*

Diversification: Currently 14,000 mushroom species are known to exist. Of these, about 50% or 7,000 species are considered to have varying degrees of edibility and almost 3,000 species from 31 genera are regarded as prime edible mushroom. To date only 200 of them are experimentally grown, 100 of them economically cultivated, approximately 40 commercially cultivated and about 10 have reached to industrial scale production in many countries. Furthermore, about 2,000 are medicinal mushrooms with varieties of health attributes. Different mushrooms require different temperatures. It is therefore, very important to go for diversification. High temperature tolerant varieties should be introduced. Varieties suitable for cultivation in India are listed in Table 16.2.

Table 16.2. Varieties suitable for cultivation in India

Mushroom variety	*Temperature range (°C)*
Oyster mushroom (*Pleurotus* species)	20–32
Milky mushroom (*Calocybe indica*)	25–35
Wood ear mushroom (*Auricularia polytricha*)	22–30
Temperature tolerant Button mushroom (*Agaricus bitorquis*)	25–30
Paddy straw mushroom (*Volvariella volvacea*)	25–35
Reishi (*Ganoderma lucidum*)	28–35
Black poplar mushroom (*Agrocybe aegerita*)	24–30

Integrated pests and disease management: Incidence of various insect pests is increasing particularly in the most commonly cultivated white button mushroom. Work on integrated disease and pest management should be carried out especially on high temperature requiring strains or species. Evolve resistant strains for various biotic and abiotic stresses. Mushroom species like *Lentinula edodes* and *Flammulina velutipes* should be introduced which grow almost at the same climate as *A. bisporus* but incidence of various insect pests and diseases is very low.

Waste management through mushroom cultivation: Potentials of mushrooms as a livelihood generator and the prospects for developing protein rich-food, pharmaceuticals and agro-waste management through mushroom cultivation needs to be tapped by creating awareness about nutritional and medicinal values of mushrooms so that per capita consumption increases. The 1 billion plus population of the country it self is a big market. If even 1% of the agro-waste is utilized, India can produce about 3.0 million tonnes of mushroom per annum, i.e., about 20% of the world production and, about 15 million tonnes of organic manure/ vermicompost generating employment to millions. Spent mushroom substrate (SMS) has its own unique traits as organic manure. SMS adsorbs the organic and inorganic pollutants and harbours diverse category of microbes having capabilities of biological breaking down the organic xenobiotic compounds. Use of spent mushroom substrate for biogas (energy) generation has several potentials and multiple benefits too.

Mitigation Measures of Reduce Emission of CO_2/GHG and Carbon Sequestration Potential of Crops

Some of the mushrooms require compost for their cultivation. During the composting some gases are released in the atmosphere. Efforts should be made to reduce the composting period so that the emission of these gases could be brought to the minimum by the use of thermophillic microorganisms. Alternatively, shift the cultivation to those mushrooms which do not require the composted substrate, e.g., *Pleurotus* spp, *Calocybe indica, Auricularia polytricha, Ganoderma lucidum, Lentinula edodes* and *Flammulina velutipes.* Higher CO_2 levels promote the growth of these mushrooms during the spawn run period.

FUTURE STRATEGIES FOR RESEARCH

- Efforts should be made to preserve the existence and distribution of fauna and flora.
- Development of short duration varieties should be undertaken.
- Diversification in mushroom cultivation should be encouraged.
- Evolve high temperature resistant strains in *Agaricus bisporus* and introduce *Agaricus bitorquis* (25–30 °C) a high temperature requiring sister species of button mushroom.
- Work on integrated disease and pest management should be carried out.
- More and more agricultural waste should be used for mushroom production.

CONCLUSION

Mushroom cultivation is mostly an indoor activity under the controlled environmental conditions, therefore is less prone to the direct effects of changing climatic conditions like elevated CO_2 and rise in temperature. However, indirect effects of global warming are affecting mushroom cultivation in several ways. Mushroom biodiversity is facing extinction due to continuing decline in their natural habitat due to increasing frequency of forest fires attributed to global warming. Growers are facing more problems abnormalities like stroma formation and diseases in years with higher temperature in the months of October and November limiting the production period of white button mushroom under natural conditions (seasonal crop). Under controlled environmental conditions, the production cost is increasing. Growers are compelled to grow-low value mushroom species. Mushroom cultivation has potential for management of agricultural wastes by using it as substrates and alternative to the burning of these wastes, thus restricting environmental pollution. Preservation of exiting fauna and flora, reduction of composting periods, integrated pest management techniques and heat tolerant mushroom species to save on energy and operating costs in mushroom cultivation under controlled conditions are the desired research and development activities.

References

Houghton, J.T., Jenkins, G.J. and Ephramus, J.J. 1990. Climate change: The IPCC Scientific Assessment. Cambridge University Press, Cambridge.

Kumar Satish and Sharma, S.R. 2001. Studies on seasonal abundance of mushroom pests. *Mushroom Research* **10**:121-23.

17

Climate Change and its Impact on Orchids

R. P. Medhi, D. Barman and B. Rai

CLIMATE change is a movement in climate system because of internal changes within the climate system or in the interaction of its components or because of changes in external forcing either by natural factors or anthropogenic activities. The fluctuation of climate is not random phenomenon but organized events which are controlled by forces or energy resources either associated with the earth itself or with the planetary bodies of the solar system. Some of the changes are induced by the human activities. The release of greenhouse gases (CO_2, water vapour CH_4 and NO_2) in atmosphere mainly by human intervention in recent years is mainly attributed for climate change. Based on current emission levels, it is believed that the global temperature will rise between 1 and 3.5 °C between now and the year 2100. These changes may have serious impacts on production of many of our agricultural and horticultural crop species.

Orchids form 9% of our flora and are the largest botanical family of higher plants in India. It is estimated that about 1,300 species (140 genera) of orchids are found in our country with Himalayas as their main home and others scattered in Eastern and Western Ghats. Orchids occupy top position among all the flowering plants valued for cut flower production and as potted plants. They are known for their longer lasting and bewitchingly beautiful flowers, which fetch a very high price in the international market. Mountains are a home to majority of orchids, climate change can have serious impacts to mountain biodiversity causing retreat or even disappearances of certain orchid species.

IMPACT ANALYSIS ON ORCHIDS

With over 25,000–30,000 species of orchids (largest group of angiosperms) in the world, India's contribution to the existing germplasm is 1,300 species scattered over various lowlying and mountainous regions. Orchids are categorized as epiphytes, lithophytes and terrestrials based on their growth habitats and there can be temperate, tropical and alpine orchids based on their requirement of light and temperature. Orchids till very recent decades were grown mostly as hobbies. Although orchids are considered to be hardy plants having the capability to withstand short spells of tough climatic variations, but the impact of elevated CO_2 level of 550 ppm and 1 °C rise in temperature over the current climate may have consequences, which has been summarized here.

Elevated CO_2 and Temperature

Productivity

Most orchids follow Crassulacean Acid Metabolism (CAM). The CAM is a mechanism whereby orchids typically take up and store carbon dioxide during the night and use it in photosynthetic carbon dioxide fixation during the day, when sunlight is available. Doubling the carbon dioxide concentration will cause stomatal conductance of leaves to decrease by about 40% while the whole crop cumulative transpiration over the course of season may only be reduced by 0–12%. Increase in water-use efficiency of CO_2 enriched crops is largely due to sizable increases in photosynthesis, growth and yield.

A survey carried out by the Botanical Society of the British Isles (BSBI) in association with plant life, a wild-flower charity reported that the bee orchid [*Ophrys apifera*] and the pyramidal orchid [*Anacamptis pyramidalis*] have virtually doubled in frequency since 1987. The cause for this doubling of orchid numbers has been attributed by the society to global warming.

***Effect on natural pollinators*:** Most orchids have adapted over time to natural variations between day and night temperatures. Typical changes in temperature will trigger an orchid to produce flowers. Secondly an orchid will flower at the time of the year in which a particular insect species is active, so that its flowers can be pollinated. Increase in temperature with increased atmospheric CO_2 levels although results in vegetative growth of the plant, flowering period would be adversely affected. Early or late flowering of any particular orchid species would indirectly affect its pollination. Most slipper orchids are pollinated by various bee species, which are active generally in the spring and summer. Unscheduled flowering in terms with its pollinator's activity would largely affect the species successive survival.

***Effect on blooming of Orchids*:** Temperature is the crucial factor in blooming *orchids*. *Cymbidium* prefers cold temperature and do not flower in warm areas. Minimum night temperature can also be as low as 7.2–10 °C and the maximum day temperature should be below 30 °C and 21 °C–26.5 °C is considered optimum for most of the *Cymbidium*. Night temperature between 10 °C to 13 °C is favorable for the initiation of flower. In extreme condition, *i.e.* too high and too low temperature, plant may survive but the flowers will not produce. Generally, the ideal day temperature for *Dendrobium* is 23 °C–29 °C and ideal night temperature is 15 °C–18 °C. The *Phalaenopsis* and its hybrids require an ideal day temperature of 27 °C–30 °C and low of 22 °C. The *Phalaenopsis* can tolerate shorter periods of high temperature, i.e., greater than 35 °C. A difference between the day and night temperatures is less than 20 °C is tolerable but with such temperatures over extended period the plants tend to become dehydrated and their leaves become limp and more flexible.

Quality

The post-harvest life of orchids depends on genetic, environmental factors including effects or intensity and duration of light, temperature, relative humidity and composition and pressure of air and most importantly total radiant energy received by the plant. The temperature and light intensity directly influenced the levels of photosynthates in flowers. The photosynthate levels in flower tissues increase as the radiant flux density increases, unless excessive temperature causes a net reduction in light due to climatological change which influences the post-harvest life of orchids. Drastic change in temperature during the flowering period will lead to aborted flower or even will not flower. Increasing temperature during flowering period reduced post-harvest life of *Cymbidium*. The post-harvest life was found to be reduced by 7–10 days. Furthermore, increased temperature over longer period results in deformed flower or shorted stem length.

Pests of Target Crop

Warmer temperature directly increases both vector and parasitic reproduction and development, rates at which pathogens are acquired, despite shortening mean daily survivorship (Patz *et al*., 2000; Liang *et al*., 2002). Orchids are subject to attack by various kinds of pest like scale, mite, aphid, shoot borer, mealy bug, black weevil, grass hopper, slug and snails *etc*. Moderate temperatures coupled with high rainfall enhance the insect population. Insect abundance varies with seasonal variations in climate. Aphids of orchids are abundant during May-June, whereas grasshopper, scale insects mealy bug, slug and snails are on July-August when there is rain. Temperature also affects behavior of insects. Some of the insects of orchids remain totally inactive at both high and low temperature or more actively along a temperature gradient until a preferred zone is encountered. In orchids, black rot, anthracnose, wild, petal blight, *Cymbidium* mosaic virus and Odontoglossum ring virus are common. Like insects, diseases of orchids depends on temperature range, rainfall, wind for the development rate, survival, phenology, distribution, continuity of population, migration and their establishment and out break.

Other Indirect Effects

***Habitat destruction*:** Most epiphytes, lithophytes and terrestrial orchids own their native habitats to the forest areas. Huge forest trees provide orchids the required shade form sunlight and rainfall. Rates of deforestation and forest degradation have greatly increased the world over. The impact of rising temperature and rising carbon dioxide on forests needs special attention because a very large population of orchid species thrives in the forest areas. If increasing atmospheric carbon dioxide levels and deforestation lead to hotter, dryer weather as predicted (Costa and Foley, 2000; Zhang *et al*., 2001; De Fries *et al*., 2002) then forests will burn more frequently. Destruction of trees would severely affect orchid habitats.

***Migration of Orchids*:** Orchids species based on their light and temperature requirements are categorized as tropical, temperate and alpine. Although orchids because of its wild hardy nature withstand slight variations of temperature and light, reports indicate that climatic variations over a decade has forced orchids to migrate towards north in search for cooler climate. Orchids have tiny seeds and their seeds can be carried long distances.

Aberrant Weather and Extreme Events

- Cymbidium prefers an optimum temperature of 21 to 26.5 °C (maximum) and minimum temperature of 10–13 °C for growth and flowering. Due to variation in weather conditions during the last few years, Cymbidiums showed poor flowering in many places where it usually flowered well. The same types of observations were also recorded in Cattleya, Phalaenopsis and Oncidium hybrids.

- Drastic changes in temperature during the flowering period at times aborted the flower of Cymbidium and other orchids and also reduced the keeping quality.
- Sometimes, chilling injury was also observed in Cattleya and Cymbidium with the symptom of browning of the throat and blackening of the column.
- Shorter stems and deformed flowers were produced due to higher temperature in orchid.
- Orchids are subjected to attack by various kind of pests like scale mite, aphid, shoot borer, mealy bug, black weevil, grasshopper, slug and snails. However, during higher temperatures coupled with rainfall insects like aphid, scale, mite, shoot borer, grass hopper, slug and snail were more prevalent. It was further observed that higher rainfall coupled with a temperature range of 24–30 °C and foggy weather invited black rot, anthracnose diseases. At extreme rainfall sclerotia was observed. Bloom blight infected during winter when the temperature suddenly changed.

Adaptation Measures for Climate Change and Global Warming

Maintaining and restoring native ecosystem: The growth conditions required for orchids are very different in comparison to other horticultural crops and re-creation of nature identical growth conditions is although not impossible but very difficult. Thus, safeguarding orchids in lieu of changing climatic conditions would be an extensive follow up of processes that would promote maintenance and restoration of native ecosystems. Orchids can be tropical, temperate or alpine depending on its environmental requirements. Existing status of ecosystems could be enhanced by massive afforestation and reforestation activities, inclusion of new species of plant and strict measures to prevent loss of gene pool by any natural or unnatural activities.

Attention to indigenous traditional knowledge (ITK): The ITK is employed to mean knowledge, innovations and practices of indigenous and local communities embodying traditional life-styles; the wisdom developed over many generations of holistic traditional scientific utilization of lands, natural resources, and environment. It is generally passed down by word of mouth, from generation to generation and is, for the most part, undocumented. The same needs to be documented and disseminated to help the conservation of orchids particularly medicinal orchids biodiversity.

Managing habitat for rare, threatened and endangered species: There are over 25,000 different types of orchids and many of them are either threatened, endangered or extinct usually due to habitat destruction or poaching. Uncontrolled grazing and pesticide use also threatens orchids. Moreover, the very laws like CITES designed to protect these plants are also threatening their future says many researchers. Orchid gardens, orchid sanctuaries and marked forest reserves are easy and less expensive means of conserving these rare orchids. While cryo-preservation although expensive can be considered an alternative means. Community awareness and training also to a large extent goes very favorable in the effective conservation of orchids.

Growing orchids under elevated CO_2: Elevated CO_2 levels with increase in temperature increase the vegetative growth of the plant. (*i.e.* significantly shorten production time). For C3 orchids, CO_2 enrichment should commence at sunrise or when photoperiod begins and recommended level is 700–1500 ppm. For CAM orchids CO_2 piping should commence before sunset and continued though darkness hours and stop when photoperiod begins. However, this process would be most beneficial if followed in the lines of carbon capture and storage (CCS) technology.The CCS technology strips CO_2 out of exhaust gases and stores it underground.

Selection of orchid media: Orchid media may consist of coconut husk, leaf mould, charcoal, brick pieces, perlite, vermiculite, and tree bark. Although the media ingredients remains same for most orchids species what differ is the ratio in which the various components are mixed to achieve maximum growth maintaining high water efficiency use. In view of high temperatures and increasing CO_2 levels orchid media with excellent water absorption capacity would be a boon to the orchid growers.

Use of shelters against intense heat: Each orchid is highly demanding in terms of its growth environments. Temperate orchids like *Cymbidiums* would not tolerate increases in temperature over long durations. Their survivability would be exhibited in abnormal morphological features like shortened stem, yellowing of leaves, deformed flower *etc*. Thus to cope with changing climate which is by and large characterized by high temperatures and elevated CO_2 levels, cool loving orchids are usually protected by shading them. Agronets of varied transmission percentages (30, 50 and 75%) are commercially available in the market. Thus depending upon the intensity of the sunlight required shade percentage may be used to protect the plants. Orchids are grown in greenhouses, thus the presence of greenhouse effect cannot be neglected. Thus construction of a greenhouse — a polyhouse or a glasshouse — depends on its location and the climatic and environmental factors surrounding it. Thus, areas experiencing high degree of sun's radiation should have houses with proper ventillation and at extreme cases the house should be fitted with fans or air conditioners and proper fogging and misting systems to provide the required temperature and humidity for optimum growth of the plant.

Improved irrigation practices: Orchid media is usually porous so a small amount of water generally suffices the need of water for the plant. Epiphytes are naturally adjusted to utilizing moisture from air however, when growing under greenhouse conditions care is needed to be taken. The basic problem of water use arises during winter season when the natural water canals runs dry in the mountains. To overcome these circumstances of water shortage, rain water harvesting during rainy season can be proven as a good bait. Sprinkler irrigation and drip irrigation are effective and proven techniques for achieving sizeable economies in irrigation water because this system other than promoting effective watering of orchid media also maintains humidity in the greenhouse.

Mitigation Measures to Reduce Emission CO_2/GHG and Carbon Sequestration Potential of Crops

Although mitigation measures are planned and executed on a much higher platform, small contributions would only add up to the much larger benefits flowing out of it. Considering the effectiveness with which orchids grows in wild habitat, mitigation measures like afforestration, reforestation, improved forest and agroforestry, not only increase carbon dioxide uptake, but also results in improved habitat conservation of orchids. CCS is an approach to mitigating global warming based on capturing CO_2 from large point sources. Most orchids follow CAM.

These plants open their stomatas during the night and close them during the day. However, if humidity is high and the plants are watered correctly, the orchids that normally close their stomatas during the day will keep them open thus enabling the plant to take up CO_2 during the day. However, carbon sequestration potential of orchids needs to be assessed.

FUTURE STRATEGIES FOR RESEARCH

- Develop micropropagation technology to conserve threatened orchids.
- Develop conservational strategies for assisted migration of orchids, if needed especially in the rocky mountainous regions of the country.
- Breeding orchids relatively tolerant to abiotic and biotic stresses.
- Development and validation of orchid crop model for tropical conditions and species.
- Precise impact assessment of global warming on orchids in India through simulation studies with appropriate crop models.

CONCLUSION

Orchids are ornamental plants of high economic value in international and domestic markets. They are valued for cut flower production and as potted plants. In India, orchids with great biodiversity comprising 1,300 species in as many as 140 genera mainly inhabit the Himalayas and are scattered in mountainous regions of Eastern and Western Ghats. Orchids prefer cold temperature and do not flower in warm environs. High temperature (>26 °C) affects blooming and may lead to flower drop or no flowering at all and the keeping quality of flowers is also reduced by 7–10 days. The climate change and global warming will affect the already fragile ecosystems of these mountainous regions with serious threat to biodiversity of orchids. However, some species like bee orchid (*Ophrys apifera*) and pyramidal orchid (*Anacamptis pyramidalis*) may be benefited doubling their population. The challenge is to protect the biodiversity and habitat of orchids through biotechnology, afforestation and assisted migration of species to new suitable environments if needed. The information on response of orchids to rise in temperature and elevated CO_2 is scanty. Accurate impact analysis of climate change and global warming through crop simulation and modeling and breeding of heat and drought tolerant orchids along with developing improved cultivation techniques would be the research priorities for orchids.

References

Costa, M.H. and Foley, J.A. 2000. Combined effects of deforestation and double atmospheric CO_2 concentrations on the climate of Amazonia. *Journal of Climate* **13**:18-34.

DeFries, R.S., Bounoua, L. and Collatz, G.J. 2002. Human modification of the landscape and surface climate in the next fifty years. *Global Change Biology* **8**:438-58.

Liang, S.Y., Linthicum, K.J. and Gaydos, J.C. 2002. Climate change and the monitoring of vector borne diseases. *Journal of the American Medical Association* **287**:2286.

Patz, J.A., Graczyk, T.K., Geller, N. and Vittor, A.Y. 2000. Effects of environmental change on emerging parasitic diseases. *International Journal for Parasitology* **30**:1395-1405.

Zhang, H., Henderson-Sellers, A. and McGuffie, K. 2001. The compounding effects of tropical deforestration and greenhouse warming on climate. *Climate Change* **49**:309-38.

18

Climate Change and its Impact on Medicinal and Aromatic Plants

Manish Das

MEDICINAL and aromatic plants (MAPs) are a diverse group of plant species. They include annuals, perennial herbs, shrubs and trees. Usually all plant parts of MAPs are used and they grow in all habitats as well as in all climatic regions of the world. About 50,000–70,000 plant species are used in traditional and modern medicines throughout the world. These species make an essential contribution to healthcare, provide an important source of income to rural and tribal harvesters. They are utilised for their aromatic properties, as raw material in industry for herbal products, pharmaceuticals and fragrances. Majority of MAPs are collected from the wild.

There is resurgence in the interest and use of medicinal plant products in India. Interest in "nutraceuticals" (functional foods) containing phytochemical constituents that can have long-term health promoting or medicinal qualities is growing. Horticultural research on medicinal plants has focused on enhancing optimal growth and production in cultivation. This has been especially pertinent as many medicinal plants are still harvested in the wild, and conditions for growth in cultivation have not been optimized (Briskin, 2000). Many of the world's poor depend directly on harvesting non-timber forest products, edible, medicinal and aromatic plants for livelihoods and sustenance (Ziska *et al.,* 1996).

In developing countries, World Health Organization (WHO) reported that 3.5 billion people rely on plants as components of their primary healthcare. Even in developed countries, where synthetic drugs dominate, 25% of all prescriptions dispensed from community pharmacies from 1959 through 1980 contained plant extracts or active principles prepared from higher plants.

Studies on impact of climate change on medicinal and aromatic plants is severely lacking except some sporadic work under controlled environment in a few crops. Elevated CO_2 has been demonstrated to increase the biomass of plants grown for medicinal purposes, with simultaneous increase in the concentrations of disease-fighting substances produced within them. However, until recently, a few studies had investigated the effects of atmospheric CO_2 enrichment on specific plant compounds of direct medicinal value.

Several chemicals derived from medicinal and aromatic plants are historically acknowledged to have pharmaceutical value (Table 18.1).

Table 18.1. Plant-derived pharmaceutical drugs and their clinical usage.

Drug	*Action/clinical use*	*Species*
Acetyldigoxin	Cardiotonic	*Digitalis lanata*
Allyl isothiocyanate	Rubefacient	*Brassica nigra*
Atropine	Anticholinergic	*Atropa belladonna*
Berberine	Bacillary dysentery	*Berberis vulgaris*
Codeine	Analgesic, antitussive	*Papaver somniferum*
Danthron	Laxative	*Cassia spp.*
L-Dopa	Anti-Parkinson's	*Mucuna spp.*
Digitoxin	Cardiotonic	*Digitalis purpurea*
Ephedrine	Antihistamine	*Ephedra sinica*
Galanthamine	Cholinesterase inhibitor	*Lycoris squamigera*
Kawain	Tranquilizer	*Piper methysticum*
Lapachol	Anticancer, antitumor	*Tabebuia spp.*
Ouabain	Cardiotonic	*Strophantus gratus*
Quinine	Antimalarial	*Cinchona ledgeriana*
Salicin	Analgesic	*Salix alba*
Taxol	Antitumor	*Podophyllum peltatum*
Vasicine	Cerebral stimulant	*Vinca minor*
Vincristine	Antileukemic agent	*Catharanthus roseus*

Source: Ziska *et al.*, 2005

IMPACT ANALYSIS ON MEDICINAL AND AROMATIC PLANTS

Effect of elevated CO_2

In controlled environments, experiments have shown beneficial effects of elevated CO_2 on productivity and quality of various products and constituents of medicinal and aromatic plants.

Productivity

In sour orange crop the atmospheric CO_2 enrichment by 75% increased total fruit production by about 80% (Idso *et al.*, 2001). In the woolly foxglove (*Digitalis lanata*) plant, producing the cardiac glycoside digoxin used in treatment of cardiac insufficiency, tripling of the air's CO_2 content in phytotron increased dry weight of woolly foxglove by 63% and 83% under well-watered and water-stressed conditions, respectively (Stuhlfauth *et al.*, 1987). In a subsequent whole-season field experiment, tripling of the air's CO_2 concentration increased plant dry weight/unit by 75% (Stuhlfauth and Fock,1990).

Elevating air's CO_2 concentration by 75% increased spider lily's belowground bulb biomass containing disease-fighting substances by 56% (Idso *et al.*, 2000).

Ultra-high CO_2 levels (3,000 µl CO_2/liter) increased fresh leaf, root and shoot weight of spearmint (*Mentha spicata* L.), thyme (*Thymus vulgaris* L.) and water mint (*Mentha aquatica* L.)

compared to high CO_2 levels (7,500 µl CO_2/litre) and ambient air (Tisserat, 2002). Pretreated shoot cultures with elevated CO_2 (10,000 µl CO_2/litre) invariably exhibited higher rates of growth and morphogenesis on transfer to soil compared to shoots pretreated with ambient air, regardless if subsequently grown in ambient or elevated CO_2 levels.

Elevated CO_2 significantly increased growth and net photosynthetic rate of *Eucalyptus camaldulensis* plantlets *in vitro*, as well as the growth and survival percentage of plantlets *ex vitro* regardless of the type of supporting materials (Kirdmanee *et al.*, 1995).

Elevated CO_2 (3,000 ml CO_2/litre of air) increased fresh weight and leaf and root numbers in cultures of lemon basil (*Ocimum basilicum*), oregano (*Origanum vulgare*), peppermint (*Mentha piperita*), spearmint (*Mentha spicata*), and thyme (*Thymus vulgaris*) shoots compared with cultures grown on the same basal media under ambient air (Tisserat, 2002).

Quality

Protein content: Effect of atmospheric CO_2 enrichment on plant constituents of significance to human health varied from increase, decrease or no effect on protein contents of various foods (Idso and Idso 2001; Wittwer, 1995; Pleijel *et al.*, 1999). However, negative impact of atmospheric CO_2 enrichment on grain protein concentration in most cases was alleviated by higher applications of nitrogen fertilizers (Pleijel *et al.,* 1999; Kimball *et al.,* 2001).

Antioxidant content: Tripling atmospheric CO_2 concentration increased vitamin C an antioxidants of great importance to human health in tomato fruits modestly by 7% only (Barbale 1970; Madsen, 1971, 1975). However in the same species, the CO_2 enrichment of similar magnitude did not influence vitamin C, but stimulated production of vitamin A (Kimball and Mitchell, 1981).

In sour orange crop atmospheric CO_2 enrichment by 75% increased sour orange juice vitamin C concentration by 5% (Idso *et al.,* 2001). In some years when CO_2-induced increase in sour orange fruit production was much greater (3.6 times) compared to the control, the increase in vitamin C concentration was 15% (Idso *et al.,* 2001). Since, vitamin C intake correlates strongly with the consumption of citrus juice (Dennison *et al.*, 1998) in any quantity by the children (Hampl *et al.*, 1999), the role played by the rise in the atmospheric CO_2 content in increasing the vitamin C concentration of orange juice could prove to be of considerable significance to public health.

In strawberry plants (Fragaria × ananassa Duchesne cv. Honeoye) grown in clear-acrylic open-top chambers elevated levels of atmospheric CO_2 of +300 and +600 ppm had positive effect on plant antioxidant production of vitamins and minerals, ascorbic acid and glutathione (GSH), anthocyanins, flavonoids, and phenolic acids in ripe fruits (Wang *et al.*, 1996, 2003; Heinonen *et al.*, 1998). Strawberries have shown a remarkably high scavenging activity toward chemically generated radicals, thus making them effective in inhibiting oxidation of human low-density lipoproteins (Heinonen *et al.*, 1998). Studies have shown that strawberries had high oxygen radical absorbance activity against peroxyl radicals, superoxide radicals, hydrogen peroxide, hydroxyl radicals, and singlet oxygen. (Wang and Jiao, 2000; Wang and Lin, 2000). Thus, it may be concluded that atmospheric CO_2 enrichment could make a good thing (strawberries) even better.

***Medicinal Constituents*:** Primitive medical records indicate that extracts from many species of plants have been used for treating a variety of human ailments (Machlin, 1992; Pettit *et al.*,

1993, 1995). In modern times the practice has continued, with numerous chemotherapeutic agents being isolated (Gabrielsen *et al.,* 1992a). Many of the chemicals derived from medicinal and aromatic plants are historically acknowledged as having pharmaceutical value (Table 18.1). However, until recently the effects of atmospheric CO_2 enrichment on specific plant compounds of direct medicinal value had not been investigated. Stuhlfauth *et al.* (1987) studied the individual and combined effects of atmospheric CO_2 enrichment and water stress on production of secondary metabolites. In the woolly foxglove (*Digitalis lanata*) plant, producing the cardiac glycoside digoxin used in the treatment of cardiac insufficiency, tripling of the air's CO_2 content in a phytotron increased the concentration of digoxin by 11% under well-watered and 14% under water-stress conditions (Stuhlfauth *et al.*, 1987). In a subsequent whole-season field experiment, tripling of the air's CO_2 concentration increased plant dry weight/unit by 75% and digoxin yield/unit dry weight by 15% resulting in doubling of total digoxin yield/ha (Stuhlfauth and Fock,1990).

The spider lily (*Hymenocallis littoralis* Jacq. Salisb.) is known, since ancient times to possess anti-tumor properties, that are effective against lymphocytic leukemia and ovary sarcoma (Pettit *et al.*, 1986) and it exhibits strong anti-viral activity against Japanese encephalitis and yellow, dengue, Punta Tora and Rift Valley fevers (Gabrielsen *et al.,* 1992 a,b). Elevating air's CO_2 concentration by 75% increased spider lily's belowground bulb biomass containing disease-fighting substances by 56% (Idso *et al.*, 2000). In addition, for these specific substances, a 6% increase in the concentration of a two-constituent (1:1) mixture of 7-deoxynarciclasine and 7-deoxy-trans-dihydronarciclasine, an 8% increase in pancratistatin, an 8% increase in trans-dihydronarciclasine, and a 28% increase in narciclasine were also recorded. Considering the 56% increase in bulb biomass the total mean active ingredient increased by 75% (approximately) due CO_2 enrichment (Idso *et al.*, 2000).

Pest of Target Crop

The disruption of relationships between MAPs and their pollinator or seed disperser under the impact of climate change is one of the biggest concerns (Idso *et al.*, 2007).

INDIRECT EFFECT OF CLIMATE CHANGE

***Extinction of medicinal and aromatic plants*:** Different plant species will respond differently to climate change and so as medicinal plant species. Some species will stay in place but adapt to new climatic conditions through selection or plasticity. Other species will move to higher latitudes or altitudes. Some species particularly the medicinal and aromatic plants may become extinct (Keutgen *et al.*, 1997). One of the biggest concerns of this community reshuffling is the disruption of food webs and coevolved mutualisms, such as the relationships between a plant and its pollinator or seed disperser (Idso *et al.*, 2007). If species that rely on each other no longer co-occur in the same time or space, both may be driven to extinction. Diseases, pests, and invasive species may spread into new ranges putting more pressure on fragile communities including MAPs. Maintaining biodiverse communities will become an even greater conservation priority and challenge. The 'weedy' species with fast generation times and wide ecological tolerances are more likely to adapt or migrate quickly and are more likely to flourish. Conservative species with specific habitat requirements or long generation times are more prone to the threat of extinction. At present an estimated one-quarter of vascular plant species are under threat in the wild (Idso *et al.*, 2007).

Aberrant Weather and Extreme Events

Information on effect of aberrant and extreme climatic events on medicinal and aromatic plants is lacking due to non availability of proper records and reports.

Adaptation Measures for Climate Change and Global Warming

- Future vulnerabilities of medicinal and aromatic plants to climate change can be reduced by conservation of endangered flora and fauna. These will include cultivation of medicinal and aromatic plants.
- Maintaining genetic diversity in natural ecosystem is key to adaptation of MAPs.
- Traditional indigenous knowledge that can avert climate change impacts should be examined and documented, while promoting traditional art and craft based livelihoods through involvement of local people and income generation through eco-tourism.

Mitigation Measures to Reduce Emission of CO_2/GHG and Carbon Sequestration Potential of Crops

The MAPs of perennial nature like trees and shrubs may have valuable mitigation potential through carbon sequestration. Proper assessment of their potential is lacking. Other mitigation measures are as follows:

- Resource conservation techniques.
- Organic farming.

FUTURE STRATEGIES FOR RESEARCH

- Lack of long-term phenological records except for a few species hampers the understanding of the effects of climate change on medicinal plants. All available information needs to be compiled and analysed.
- Studies on physiological response of medicinal plants like senna, ashwagandha, tinospora, safed musli, aloe vera, isabgol, lemongrass, palmarosa, kalmegh, patchouli, guggal etc. and their phytomedicinal chemical production with respect to climate change and weather aberration.
- Identification of valuable wild plant populations most likely to be affected by climate change for their effective conservation.
- Identify the strategic approaches, priority interventions and partnerships necessary for taking full advantage of herbs and medicinal and aromatic plants experiments.
- Experiments in open top chambers and controlled environments to simulate the natural condition and various biotic and abiotic stresses under climate change.
- Indigenous knowledge of herbal, medicinal and aromatic plants need to be compiled and documented for their conservation and possible cultivation.
- Models are useful tools for predicting responses of single species, multi-species assemblages, global vegetation patterns, to climate change for delineating hardiness zones for medicinal plants. These models need to be developed urgently.

CONCLUSION

Medicinal and aromatic plants (MAPs) are used in traditional and modern medicine providing an important source of income to rural and tribal harvesters. They are used for their aromatic properties, as raw material in industry for herbal products, pharmaceuticals and fragrances. There is a growing interest in these plant species for various reasons. Horticultural research on medicinal plants has focused on developing technology for cultivation, because majority of medicinal plants are still harvested in the wild and their true potential for mankind is yet to be realized.

Knowledge on the impact of climate change on MAPs is lacking except some sporadic work under controlled environment demonstrating increase in the biomass of these plants with simultaneous increase in concentrations of disease-fighting substances in them under elevated CO_2. However, little is known about the effect of rise in temperature on growth yield and quality of their products. Extinction of these valuable species due to disruption of their natural habitat under global warming and climate change is a major concern. Therefore, conservation of genetic biodiversity of these species becomes first priority possibly by cultivation and other means. Traditional indigenous knowledge can be helpful in minimizing the impact of climate change. Compilation and documentation of this knowledge, while promoting traditional art and craft based livelihoods through involvement of local people and income generation through eco-tourism is required. Research on effect of rise in temperature on MAPs needs to undertaken urgently along with development and application of modern approaches of modeling for predicting responses of single species, multi-species assemblages, global vegetation patterns, to climate change for delineating hardiness zones and new areas for relocation of these plants.

References

Barbale, D. 1970. The influence of the carbon dioxide on the yield and quality of cucumber and tomato in the covered areas. *Augsne un Raza* (Riga) **16**:66-73.

Briskin Donald, P. 2000. Medicinal Plants and Phytomedicines. Linking Plant Biochemistry and Physiology to Human Health, *Plant Physiol.* **124**:507-14.

Dennison, B.A., Rockwell, H.L. and Baker, S.L. 1998. Fruit and vegetable intake in young children. *J. Amer. Coll. Nutr.* **17**:371-78.

Gabrielsen, B., Monath, T.P., Huggins, J.W., Kefauver, D.F., Pettit, G.R., Groszek, G., Hollingshead, M., Kirsi, J.J., Shannon, W.F., Schubert, E.M., Dare, J., Ugarkar, B., Ussery, M.A. and Phelan, M.J. 1992a. Antiviral (RNA) activity of selected Amaryllidaceae isoquinoline constituents and synthesis of related substances. *Journal of Natural Products* **55**:1569-81.

Gabrielsen, B., Monath, T.P., Huggins, J.W., Kirsi, J.J., Hollingshead, M., Shannon, W.M. and Pettit, G.R. 1992b. Activity of selected Amaryllidaceae constituents and related synthetic substances against medically important RNA viruses. *In:: Natural Products as Antiviral Agents.* Chu, C.K. and Cutler, H.G. (Eds.),Plenum Press, New York, NY, pp. 121-35.

Hampl, J.S., Taylor, C.A. and Johnston, C.S. 1999. Intakes of vitamin C, vegetables and fruits: Which schoolchildren are at risk? *J. Amer. Coll. Nutr.* **18**:582- 90.

Heinonen, I.M., Meyer, A.S. and Frankel, E.N. 1998. Antioxidant activity of berry phenolics on human low-density lipoprotein and liposome oxidation. *Journal of Agricultural and Food Chemistry*, **46**:4107-12.

Idso, C.D. and Idso, K.E. 2000. Forecasting world food supplies: The impact of the rising atmospheric CO_2 concentration. *Technology* **7S**:33-55.

Idso, S.B., Craig, D. and Idso, K.E. 2007. Meme 98: Enhanced or Impaired? Human Health in a CO_2-Enriched Warmer World. Center for the Study of Carbon Dioxide and Global Change P.O. Box 25697, Tempe, AZ 85285-5697, USA www.co$_2$science.org.

Idso, S.B. and Idso, K.E. 2001. Effects of atmospheric CO_2 enrichment on plant constituents related to animal and human health. *Environmental and Experimental Botany* **45**:179-99.

Idso, S.B., Kimball, B.A., Pettit III G.R., Garner, L.C., Pettit, G.R. and Backhaus, R.A. 2000. Effects of atmospheric CO2 enrichment on the growth and development of *Hymenocallis littoralis* (Amaryllidaceae) and the concentrations of several antineoplastic and antiviral constituents of its bulbs. *American Journal of Botany* **87**:769-73.

Keutgen, N., Chen, K. and Lenz, F. 1997. Responses of strawberry leaf photosynthesis, chlorophyll fluorescence and macronutrient contents to elevated CO_2. *Journal of Plant Physiology* **150**:395-400.

Kimball, B.A. and Mitchell, S.T. 1981. Effects of CO_2 enrichment, ventilation, and nutrient concentration on the flavor and vitamin C content of tomato fruit. *Hort Science* **16**:665-66.

Machlin, L.G. 1992. Introduction. In: Sauberlich, H.E. and Machlin, L.J. (Eds.), Beyond deficiency: New views on the function and health effects of vitamins. *Annals of the New York Academy of Science* **669**:1-6.

Madsen, E. 1971. The influence of CO_2-concentration on the content of ascorbic acid in tomato leaves. *Ugeskr. Agron.* **116**:592-94.

Madsen, E. 1975. Effect of CO2 environment on growth, development, fruit production and fruit quality of tomato from a physiological viewpoint. In: P. Chouard, N. de Bilderling (Eds.), *Phytotronics in Agricultural and Horticultural Research.* Bordas, Paris, pp. 318-30.

Pettit, G.R., Pettit III G.R., Backhaus, R.A., Boyd, M.R. and Meerow, A.W. 1993. Antineoplastic agents, 256. Cell growth inhibitory isocarbostyrils from Hymenocallis. *Journal of Natural Products* **56**:1682-87.

Pettit, G.R., Pettit III GR, Groszek G, Backhaus RA, Doubek DL and Barr RJ. 1995. Antineoplastic agents, 301. An investigation of the Amaryllidaceae genus Hymenocallis. *Journal of Natural Products* **58**:756-59.

Pleijel, H., Mortensen L., Fuhrer, J., Ojanpera, K. and Danielsson, H. 1999. Grain protein accumulation in relation to grain yield of spring wheat (*Triticum aestivum* L.) grown in open-top chambers with different concentrations of ozone, carbon dioxide and water availability. *Agriculture, Ecosystems and Environment* **72**:265-70.

Stuhlfauth, T., Fock, H.P. 1990. Effect of whole season CO2 enrichment on the cultivation of a medicinal plant, Digitalis lanata. *Journal of Agronomy and Crop Science* **164**:168-73.

Stuhlfauth, T., Klug, K. and Fock, H.P. 1987. The production of secondary metabolites by Digitalis lanata during CO2 enrichment and water stress. *Phytochemistry* **26**:2735-739.

Tisserat Brent. 2002. Influence of Ultra-High Carbon Dioxide Levels on Growth and Morphogenesis of Lamiaceae Species in soil *Journal of Herbs, Spices & Medicinal Plants* **9**:81-89.

Wang, H., Cao, G. and Prior, R.L. 1996. Total antioxidant capacity of fruits. *Journal of Agricultural and Food Chemistry,* **44**:701-705.

Wang, S.Y. and Jiao, H. 2000. Scavenging capacity of berry crops on superoxide radicals, hydrogen peroxide, hydroxyl radicals, and singlet oxygen. *Journal of Agricultural and Food Chemistry* **48**:5677-684.

Wang, S.Y. and Lin, H.S. 2000. Antioxidant activity in fruit and leaves of blackberry, raspberry, and strawberry is affected by cultivar and maturity. *Journal of Agricultural and Food Chemistry* **48**:140-46.

Wang, S.Y., Bunce, J.A. and Maas, J.L. 2003. Elevated carbon dioxide increases contents of antioxidant compounds in field-grown strawberries. *Journal of Agricultural and Food Chemistry* **51**:4315-320.

Wittwer, S.H. 1995. Food, Climate, and Carbon Dioxide: The Global Environment and World Food Production. *CRC Press*, Boca Raton, FL.

Ziska, L.H. 1996. Intraspecific variation in the response of rice (*Oryza sativa* L.) to increased CO2 and temperature: growth and yield response of 17 cultivars. *Journal of Experimental Botany* **47**:1353-359.

Ziska, L.H. 2005. Alterations in the production and concentration of selected alkaloids as a function of rising atmospheric carbon dioxide and air temperature: Implications for ethnopharmacology. *Global Change Biology* **11**:1798-807.

19

Impact of Climate Change on Fungal Diseases of Horticultural Crops

P. Chowdappa

PLANT diseases cause significant losses in yield and quality of horticultural produce resulting in severe economic losses to fruit, vegetable and ornamental growers in India every year. A plant pathosystem is a system composed of host, pathogen and environment, affecting the plant-pathogen relationship and no disease can develop in the absence of conducive environment for host-pathogen interaction (Zadoks and Schein, 1979; Hardwick, 1998). Change in environmental factors, particularly temperature and leaf wetness, influence the rate of progress at any stage of plant disease development (Agrios, 1997; Coakley, *et al.*, 1999).

Historical examples of Irish potato famine (1845–47) caused by *Phytophthora infestans*, coffee rust (1860) caused by *Hemileia vastatrix* and Bengal famine (1943) attributed to rice brown spot caused by *Cochliobolus miyabeanus* remind us of the devastating effect of plant disease epidemics to equip ourselves in the context of global climate change.

The global environment is changing. Increasing concentration of greenhouse gases, particularly CO_2 in atmosphere coupled with rising temperature, altered precipitation patterns, as well as numerous other potential changes are taking place resulting in global warming and climate change (Runion, 2003). Climate change will directly impact many horticultural crops, as well as their interactions with plant pathogens influencing plant diseases and the efficacy of their management options to impact productivity and sustainability of agricultural systems (Chakraborty *et al.*,1998 and 2000b; Coakley and Scherm, 1996; Coakley *et al.*; 1999; Anderson *et al.,* 2004; Garrett *et al.,* 2006). The climate change may reduce, increase or have no effect on plant diseases. The most likely consequences are changes in the geographical distribution of host and pathogen and crop losses, caused in part by changes in the efficacy of management strategies. However, it is currently difficult to predict the precise impact on specific crop pathosystems. The relevance of climatic and biospheric change to plant diseases is receiving increasing attention worldwide (Atkinson 1993; Prestidge and Pottinger 1990; Chakraborty *et al.*,1998; Coakley 1995; Sutherst *et al.,* 1996a,b). The current knowledge on the impacts of climate change on pathogens and diseases particularly fungal diseases of horticultural crops in India is briefly reviewed here.

IMPACT ANALYSIS OF FUNGAL DISEASES ON HORTICULTURAL CROPS

Climatic factors that influence the growth, spread, and survival of crop diseases include temperature, precipitation, humidity, dew, radiation, wind speed, circulation patterns, and the occurrence of extreme events. Higher temperature and humidity and greater precipitation result in the spread of plant diseases, as wet vegetation promotes the germination of spores and the proliferation of fungi and influences the lifecycle of soil nematodes. In regions that suffer aridity, however, disease infestation lessens, although some diseases (such as powdery mildew) thrive in hot, dry conditions, as long as there is dew formation at night.

Elevated CO_2 and Temperature

***Elevated CO_2 levels*:** Limited studies on effects of CO_2 on fungal diseases indicated that disease severity increased, decreased, or remained unchanged (Chakraborty *et al.*,1998; Coakley *et al.,*1999; Thompson and Drake 1994; Manning and Tiedemann, 1995). Increased pathogen fecundity (spores produced /lesion area) at elevated CO_2 has been demonstrated in several fungi (Coakley *et al.,*1999; Hibberd *et al.*, 1996, Chakraborty and Datta 2003; Chakraborty *et al.,* 2000a). The diseases caused by pathogens that infect through stomata, such as *Phyllosticta minima,* may be reduced when the CO_2 levels increases due to reduction in stomatal opening and changes in leaf chemistry. Pangga *et al.* (2004) reported that high levels of CO_2 might prevent induced resistance as plants grow more rapidly. Elevated CO_2 causing higher carbohydrate concentration within host tissue promotes the development of some biotrophic pathogens such as rusts (Manning and Tiedemann, 1995), but inhibits others such as mildews (Hibberd *et al.*, 1996). The host resistance to early blight of tomato and potato caused by *Alternaria solani* is likely to be enhanced by the elevated CO_2 effect along with enhanced sporulation and infection through increased summer rainfall. Therefore, under climate change, damage due to early blight will depend on the balance between enhanced host plant resistance and increased pathogen sporulation.

Temperature and Relative Humidity

***Temperate fruits*:** High relative humidity and temperature between 10 and 22 °C with out free water are ideal conditions for apple powdery mildew, caused by *Podosphaera leucotricha* (Ell. & Ev.) E.S. Salmon, but an increase in summer temperatures above 30 °C inhibits infection (Hickey and Yoder, 1990). It is true for apple sooty blotch caused by *Gloeodespomigena* (Schwein.) Colby, too (Sutton, 1990). The incidence of blossom blight in stone fruits caused by *Monilinia fructicola* (Wint.) is related to temperature and duration of wetness and 3-5 hour of wetness at 20 °C can cause severe infection (Ogawa *et al.,* 1995). An increase in spring temperature may reduce the incidence of peach leaf curl, caused by *Taphrina deformans* (Burk.) Tulasne in some seasons (Pscheidt, 1995).

Tropical fruits: Downy mildew [*Plasmopara viticola* (Berk. & Curt.) Berl. & deToni], powdery mildew [*Uncinula necator* (Schwein.) Burill], anthracnose (*Elsinoe ampelina* Shear), Botrytis and other bunch rots (*Botrytis cinema* Pers. :Fr. and other species) and phomopsis

(*Phomopsis viticola* (Sacc.) are most serious diseases affecting grape (Chakraborty *et al*., 1998). Simulations of Salinari *et al.,* (2006) indicated that, warmer temperature regimes (an increase of 2-3 °C by 2080) can significantly increase the incidence of downy mildew pressure in grapes in Italy, even if precipitation decreases (76-77%). The *P. viticola* is highly dependent on wetness for both, its survival and pathogenicity. The minimum temperature for pathogen growth is 9-10 °C, while temperatures higher than 34–35 °C inhibit the pathogen influenced by weather conditions. Blaeser and Weltzien (1979) found that minimum requirement for infection is a value of 60, calculated as the product of temperature times hours of wetness at 12 °C (at least 5 hours) of wetness are necessary. Therefore, a higher temperature can compensate for reduced wetness duration. In addition, high temperature accelerates oospore germination when leaf residues are wet onto soil surface, so that less humidity can be easily compensated (Rossi *et al*., 2005). An increase in cool moist conditions in spring would favour phomopsis and black spot (*Elsinoe ampelina* Shear). Any increase in temperature, relative humidity, canopy growth, and frequency of rain events of greater extent or quantity would favour increases in levels of Botrytis bunch rots. Yellow rust (*Physopella ampelopsidis)*, confined to Bangalore blue variety earlier, is now attained to epidemic proportions on both coloured (Sharad, Flame) and colourless (Sonka and Thompson and Anabe-ei-Shahi) grapes in Bangalore and becoming a major impediment in grape cultivation.

Plantation crops: Temperatures above 30 °C or below 17 °C slackens the bacterial stripe (*Xanthomonas campestries* pv *arecae*) of arecanut (Sampath Kumar, 1992). The incidence of leaf rot of coconut (*C. gloeosporioides*) has been found to be positively correlated with rain fall and relative humidity and negatively correlated with maximum temperature and sunshine hours (Chowdappa *et al*., 2002). Local epidemic outbreaks of bud rot of coconut was known to be preceded by high rainfall during the previous months (Rillo and Paloma, 1989). About 5000 coconut palms of 25–30 years age group were killed by bud rot in Kuttiady area of Calicut district of Kerala in October 1992 due to heavy rainfall in preceding months (Chowdappa *et al*., 2002). Rain fall, number of rainy days, high RH and low sunshine hours showed positive correlation with foot rot (quick wilt) of black pepper and high temperature and sunshine hours showed negative correlation (Unnikrishnan Nair *et al.,* 1988). Outbreaks of wilt, leaf rust, and berry disease in coffee, cocoa black pod, vascular streak dieback and wilt are causing serious crop losses in India.

Vegetable crops: Verticillium wilt in tomato is sensitive to temperature. The pathogen survives as micro sclerotia in soils in the absence of a host due to limited saprophytic ability. In USA, two forms with different tolerance to high temperature have been identified. Edgington (1962) reported that the micro sclerotial strain could tolerate higher temperatures than the dark mycelial strain. Verticillium wilt in tomato crops can be controlled by soil solarization to raise the average daily maximum temperature by 10 °C at 7.5 cm depth. The fungus can be eradicated from infected stem pieces by keeping at 45 °C for less than 24 hours. Under climate change, this sensitivity to temperature may make this disease less important. Early blight of tomato and potato caused by *Alternaria solani* is very strongly influenced by physiological status of the host. Host susceptibility increases as tissue ages or is stressed due to poor nutrition or root disease. Early blight will develop when leaves remain wet for at least 3hours at 21–25 °C but conidia will germinate at temperature up to 34 °C (Madden *et al*.,1978) or when relative humidity is more than 90% (Sherf and MacNab, 1986). With each 1 °C increase in temperature in Finland, late blight of potato occurred 4–7 days earlier and host susceptibility period enhanced by 10–20 days (Kaukoranta, 1996). As result, number of sprays may increase, enhancing the cost of production and affecting the environment. Based on meta modeling approach, Andrade-Piedra *et al.* (2005) showed that late blight of potato was basically weather driven disease and changes in temperature and relative humidity are critical for disease outbreaks. Bourgeois *et al.* (2004) developed mathematical models for studying the impact of climate change on incidence of late blight of potato [*Phytophthora infestans*], apple scab [*Venturia inaequalis*], and cercospora blight of carrot [*Cercospora carotae*].

Effects of Elevated Ozone (O_3) Concentrations

Exposure of plants to O_3 may influence foliar diseases caused by fungi. Changes in the structure of leaf surfaces, *i.e.* physical topography, the chemical composition of surfaces, the structure of epicuticular wax have been reported when ozone concentration levels increased (Karnosky *et al.*, 2002). These changes in leaf structure may alter leaf surface properties such as leaf wetness and the ability of leaves to retain solutes, which will influence the ability of pathogens to attach to leaf surfaces and infect. Increased ozone levels tend to enhance the disease intensity on plants by necrotrophic fungi, root-rot fungi, and bark beetles (Sandermann, 2000). Karnosky *et al.* (2002) reported severity of rust (*Melampsora medusae* f. sp. *tremuloidae*) infections on aspen (*Populus tremuloides* Michx.) when exposed to plants to higher levels of ozone and CO_2. It was demonstrated that concentrations of O_3 at or below economic threshold (50ppb) enhanced severity of powdery mildew infection by *Sphaerotheca fuliginea* on cucumber, whereas higher levels (100 or 200 ppb) decreased the severity and also conidial germination (Khan and Khan, 1999).

Other Indirect Effects of Climate Change

Movement of pathogens: In recent years, changes in storm patterns have been reported (Emanuel, 2005; Webster *et al.*, 2005), which could influence the global movement of pathogens (Brown and Hovmoller, 2002). Long-distance dispersal of fungal spores by wind can spread plant diseases across and even between continents and reestablish diseases in areas where host plants are seasonally absent.

Host resistance: Water stress, $CO_{2,}$ elevated temperature and ozone levels have profound influence on effectiveness of host resistance. Huber and Gillespie (1992) reported that foliar pathogens had lower infections under drought situations, due to lack of required leaf wetness for spore germination. Host-pathogen gene pairs linked to resistance responded differently to different temperature as recorded in wheat-*(Puccinia recondita)* system sunflower-broomrape and rice — *Xanthomonas oryzae* pv. *oryzae* (Garret *et al.,* 2006).

Disease management: Farmers use a wide range of agricultural practices to limit losses from diseases in horticultural crops. Strategies for disease control include cultural practices (such as crop rotations), biological control, disease resistant varieties, and chemical control. The variability in CO_2 concentrations, rainfall and temperature conditions affect the optimal management of fungicidal applications (Coakley *et al.*, 1999). The biological control agents in plant disease management may fail in the field due to environmental variation and environmental extremes (Grevstad, 1999).

Aberrant Weather and Extreme Events

- The incidence of fruit rot (*Phytophthora meadii*) of arecanut was rampant during 1978 and caused losses up to 90% in India, due to changes in rainfall pattern, humidity and temperature. Favourable conditions for disease existed for a prolonged period (May—September) during 1978 with a total rainfall of 5088.6 mm (highest in the last 10 years). The rain started during the first fortnight of May itself and continued until September and the conditions of high temperature and humidity (>80 %) prevailed during the period.

Adaptation Measures for Climate Change and Global Warming

- Integrated pest management.
- Using available early warning systems for diseases.
- Biological control measures.
- Utilization of indigenous traditional knowledge base for disease control.
- Soil solarization techniques.

FUTURE STRATEGIES FOR RESEARCH

- Development and validation of weather-based disease forecasting models for Indian conditions to serve as early warning systems.
- Development of simulation model based decision support system for fungicide sprays and disease control.
- Breeding disease tolerant cultivars.
- Studies on changes in host physiology, pathogen life-cycle and host-pathogen interaction caused by changing climatic parameters.
- Assessment of impacts of extreme events, such as fire, frost, floods, etc., in addition to the impact of gradual changes in climatic factors on diseases of horticultural crops in India.
- Studies on increased pathogen fecundity at elevated CO_2 on disease resistance in crop plants.

CONCLUSION

The likely impacts of climate change on growth and yield of agricultural crops have been assessed experimentally and through simulation modeling studies, but knowledge about impact on plant diseases is still lacking. Plant pathogens may respond to climate change, for both short generation times and effective dispersal mechanisms. The elevated carbon dioxide concentrations, altered temperature and precipitation regimes under climate change will be the most important factors influencing plant diseases. Climate change and global warming may alter growth stages and/or rates of development in the life cycle and pathogenicity of pathogens, as well as modify the physiology and resistance of host plants. There is a need to understand and anticipate potential climate impacts on host-pathogen interactions, in order to evolve appropriate disease management strategies. Based on assessment of important diseases of horticultural crops in India, impacts of climate change is implicated in recent epidemic out-breaks of fruit rot (*Phytophthora meadii*) and leaf spot (*Phyllosticta arecae* and *C. gleosporioides*) of arecanut, bud rot (*Phytophthora palmivora*) of coconut, yellow rust, *Alternaria* bunch rot, *Botryodiplodia* die back and *Grenaria* leaf and fruit rot in grape, *Macrophoma* in banana, *Cercopsora* and wilt in chilli, blossom blight in mango and wilt and nodal blight in pomegranate in certain locations. Development of early warning systems for disease forecasting, adaptation techniques of integrated pest management, breeding tolerant cultivars are possible options to manage plant diseases under future climate change scenarios.

References

Agrios, G.N.1997. *Plant Pathology,* 4th edn, Academic Press, San Diego.

Andrade-Piedra, J.L., Forbes, G.A., Shtienberg, D., Grunwald, N.J., Taipe, M.V., Hijmans, R.J. and Fry, W.E. 2005. Quantification of plant disease simulation model: performance of the LATE BLIGHT model across a broad range of environments. *Phytopathology* **95**:1412-22.

Anderson, P.K., Cunningham, A.A., Patel, N.G., Morales, F.J., Epstein, P.R. and Daszak, P. 2004. Emerging infectious diseases of plants: pathogen pollution, climate change and agrotechnology drivers. *Ecol. Evol.* **19**:535-44.

Atkinson, D. 1993. Global climate change, its implication for crop protection. British Crop Protection Council Monograph No 56. BCPC, Surrey, UK.

Blaeser, M. and Weltzien, H.C. 1979. Epidemiological studies to improve the control of grapevine downy mildew (*Plasmopara viticola*). *Journal of Plant Diseases and Protection* **86:**489-98.

Bourgeois, G., Bourque, A. and Deaudelin, G. 2004. Modeling the impact of climate change on disease incidence: a bio climatic challenge. *Canadian Journal of Plant Pathology* **26**:284-90.

Brown, J.K.M. and Hovmoller, M. 2002. Aerial dispersal of pathogens on the global and continental scales and its impact on plant disease. *Science* **297:**537-41.

Chakraborty, S. and Datta, S. 2003. How will plant pathogens adapt to host resistance at elevated CO_2 under a changing climate? *New Phytol.* **159:**733-42.

Chakraborty, S., Pangga, I.B., Lupton, J., Hart, L., Room,P.M. and Yates, D. 2000a. Production and dispersal of *Colletotrichum gloeosporioides* spores on *Stylosanthes scabra* under elevated CO_2. *Environ. Pollut.* **108**:381-87.

Chakraborty, S., Murray, G.M., Magarey, P.M., Yonow, T., O'Brien, R.G., Croft, B.J., Barbetti, M.J., Sivasithamparam, K, Old, K.M., Dudzinski, M.J., Sutherst, R.W., Penrose, L.J., Archer, C. and Emmett, R.W. 1998. Potential impact of climate change on plant diseases of economic significance to Australia. *Aust. Plant Pathol.* **27:**15-35.

Chakraborty, S., Tiedemann, A.V. and Teng, P.S. 2000b. Climate change: potential impact on plant diseases. *Environ. Pollut.* **108**:317-26.

Chowdappa, P., Rohini Iyer and Gunasekaran, M. 2002 Coconut diseases. In Plant Pathology Research at Central Plantation Crops Research Institute, Kasaragod pp.17.

Coakley, S.M. 1995. Biospheric change: will it matter in plant pathology? *Canadian Journal of Plant Pathology* **17**:147-53.

Coakley, S.M. and Scherm H. 1996 . Plant disease in a changing global environment. *Aspects of Applied Biology* **45**:227-37.

Coakley, S.M., Scherm, H. and Chakraborty, S. 1999. Climate change and plant disease management. *Annu. Rev. Phytopathol* **37**:399-26.

Edgington, L.V. 1962. Influence of Connecticut temperatures on the relative pathogenicity of Maine and Connecticut *Verticillium* isolates. *American Potato Journal* **39:**261-65.

Emanuel, K. 2005. Increasing destructiveness of tropical cyclones over the past 30 years. *Nature* **436**:686-88.

Garrett, K.A., Dendy, S.P., Frank, E.E., Rouse, M.N. and Travers, S.E. 2006. Climate Change Effects on Plant Disease: Genomes to Ecosystems *Annu. Rev. Phytopathol* **44**:489-509.

Grevstad, F.S. 1999. Factors influencing the chance of population establishment: implications for release strategies in biocontrol. *Ecol. Appl.* **9**:1439-447.

Hardwick, N.V. 1998. Disease forecasting. *In: The Epidemiology of Plant Diseases*, Jonse, D.G. (Ed.), Kluwer Academic Publishers, Boston, pp. 207-230.

Hibberd, J.M, Whitbread R and Farrar JF. 1996. Effect of elevated concentrations of CO_2 in infection of barley by *Erysiphe graminis. Physiol. Mol. Plant Pathol.* **48**:37-53.

Huber, L. and Gillespie, T.J. 1992. Modeling leaf wetness in relation to plant disease epidemiology. *Annu. Rev. Phytopathol.* **30**:553-77.

Karnosky, D.F., Percy, K.E, Xiang, B., Callan, B., Noormets, A. 2002. Interacting elevated CO_2 and tropospheric O_3 predisposes aspen (*Populus tremuloides* Michx.) to infection by rust (*Melampsora medusae* f. sp. *tremuloidae*). *Glob. Change Biol.* **8:**329-38.

Kaukoranta, T.1996. Impact of global warming on potato late blight: risk, yield loss and control. *Agric, Food Sci. Finl.* **5**:311-27.

Khan, M.R. and Khan, M.J. 1999. Effects of intermittent ozone exposures on powdery mildew of cucumber. *Enviornmental and Experimental Botany* **42**:163-71.

Madden, L., Pennypacker, S.P. and MacNab, A.A. 1978. "FAST, a forecast system for *Alternaria solani* on tomato" *Phytopathology* **68**:1354-58.

Manning, W.J. and Tiedemann, A.V. 1995. Climate change: Potential effects of increased atmospheric carbon dioxide (CO_2), ozone (O_3), and ultraviolet-B (UV-B), radiation on plant diseases. *Environ. Pollut.* **88**:219-45.

Ogawa, Zehr, J.M., E.I. and Biggs, A.R. 1995. Brown rot. *In: Compendium of stone Fruit Diseases.* Ogawa, J.M., Zehr, E.I., Bird, G.W., Ritchie, D.F., Uriu, K. and Uyemoto, J.K. (Eds), American Phytopathological Society Press, St. Paul MN. pp. 7-10

Pangga, I.B., Chakraborty, S. and Yates, D. 2004. Canopy size and induced resistance in *Stylosanthes scabra* determine anthracnose severity at high CO_2. *Phytopathology* **94**:221-27.

Prestidge, R.A. and Pottinger, R.P. 1990. *The Impact of Climate Change on Pests, Diseases, Weed sand Beneficial Organisms Present in New Zealand Agricultural and Horticultural Systems. MAF* Technology, Ruakura Agricultural Centre, Hamilton, New Zealand.

Pscheidt, J.W.1995. Leaf curl. In *Compendium of Stone Fruit Diseases.* Ogawa, J.M., Zehr, E.I., Bird, G.W., Ritchie, D.F., Uriu, K. and Uyemoto, J.K. (Eds.), American Phytopathological Society Press, St. Paul MN. pp.22.

Rillo, E.P. and paloma, M.B.F. 1989. Reaction of some coconut populations and hybrids to *Phytophthora* disease in Bicol region. *Philippine Journal of Coconut Studies*, June 2-6.

Rossi, V., Caffi, T. and Melandri, M. 2005. Aggiornamenti sulla peronospora della vite. *Agronomica* **2**:38-56.

Runion, G.B. 2003. Climate change and plant pathosystems – future disease prevention starts here. *New Phytologist* **159:**531-33.

Salinari, F., Ghosue, S.W., Tubielloz Retto, R.I.F., Torioro, S.S.I.W., Panna, F., Rosenzweig, Cz., and Lodovica, M. 2006. Downy mildew (*Plasmopara viticola*) epidemics on grapevine under climate change. *Global Change Biology* **12**:1299-07.

Sampath Kumar, S.N. 1992. Bacterial leaf stripe disease of arecanut caused by *Xanthomonas campestries* pv. *Arecae*, Ph.D thesis, Indian Institute of Science, Bangalore, pp.190.

Sandermann, H. Jr. 2000. Ozone/biotic disease interactions: molecular biomarkers as a new experimental tool. *Environ. Pollut.* **108**:327-32.

Sherf, A.F. and MacNab, AA.1986. *Vegetable Diseases and their Control*, Wiley, New York.

Sutherst, R.W., Yonow, T., Chakraborty, S., O'Donnell, C. and White, N. 1996a. A generic approach to defining impacts of climate change on pests, weeds and diseases in Australasia. In *Greenhouse, Coping with Climate Change* (Eds W.J. Bouma, G.I. Pearman and M.R. Manning). CSIRO Australia. pp. 281-307.

Sutherst, R.W., Ingram, J.S.I. and Steffen, W.L. 1996b. Global Change Impacts on Pests, Diseases and Weeds Implementation Plan. GCTE Core Project Office, Canberra, Australia.

Sutton, T.B. 1990. Sooty blotch and flyspeck. *In: Compendium of Apple and Pear Diseases.* Jonesand, A.L., Aldwinckle, H.S. (Eds). American PhytopathologicalSociety Press, St. Paul MN pp. 20-22.

Thompson, G.B. and Drake, B.G. 1994. Insects and fungi on a C3 sedge and a C4 grass exposed to elevated CO_2 concentrations in open-top chambers in the field. *Plant Cell Environ.* **17**:1161-67.

Unnikrishnan Nair, P.K., Sasikumaran, S., Sukumarapillay, V. and Prasad Ral, G.S.L.H.V. 1988. Influence of weather on foot rot disease of pepper. *In: Agrometerology of Plantation Crops.* G.S.L.H.V. Prasad Rao and Nair, R.R. (Eds.), Kerala Agricultural University, Trichur.

Webster, P.J., Holland, G.J., Curry, J.A. and Chang, H.R. 2005. Changes in tropical cyclone number, duration, and intensity in a warming environment. *Science* **309**:1844-46.

Zadoks, J.C. and Schein, R.D. 1979. Epidemiology and plant disease management, Oxford University Press, New York, pp. 427.

20

Climate Change and Virus Diseases of Horticultural Crops

M. Krishna Reddy

There is a wide scientific consensus that global climate is changing as a result of human activities (IPCC, 2001). The social and economic cost of slowing the rate of climate change and responding to its impacts will be large. Climate change will have a range of positive and negative impact on agriculture. Climate variability is currently the dominant cause of short-term fluctuation in rainfed agricultural production of sub-Saharan Africa, South Asia and substantial areas of other developing regions. The greatest impacts could come from increased frequency and intensity of extreme events. The most serious form is drought, when rainfall drops substantially below the long-term mean or fails at critical points in crop development (Bruinsma, 2008). Extreme events of drought or floods can kill large expanses of trees directly by overwhelming the tree physiological capability and structural strength. The indirect impacts operate primarily through effects on resource availability, notably water resources and on ecosystems as they respond to shifts in temperature and precipitation; and through the loss of biodiversity.

Crop yields are significantly affected every year in India, and elsewhere, by diseases. The effects of climate change on crop health and, in particular, on plant viruses and vectors are not well-known. Weather and climate strongly influence plant virus diseases and vectors. The temperature and moisture conditions interacting with seasonal phenology, and stress on the host determine infection severity and distribution. Vectors transmit a number of plant viruses. Many of them are unable to establish at current climatic conditions. However, climate change is expected to influence vector establishment, spread and reproduction potential as well as transmission of viruses. Vectors established in greenhouses only up to now may escape from these and establish and damage plants outdoors. Crops might benefit from warmer temperatures and elevated levels of CO_2 in atmosphere, but may be stressed by drier soils and weather extremes under future climate scenarios. Stressed plants are often more susceptible to diseases.

Climate change could alter stages and rates of development of pathogen, modify host resistance, and result in changes in physiology of host-pathogen interactions (Coakley *et al.*, 1999). The effects of climate change would tend to be different for different pathosystems in different locations, so that generalization is a challenge (Garrett *et al.*, 2006).

IMPACT ANALYSIS OF VIRUS DISEASES ON HORTICULTURAL CROPS

Literature review provides some background on potential impact of climate change on plant diseases. Most of the literature focuses on the influence of temperature, precipitation, carbon dioxide, ozone, ultraviolet (UV) radiation on diseases and insects of agricultural crops (Coakley *et al.*, 1999; Harvell *et al.*, 2002; Manning and Von Tiedemann 1995; Sandermann 2000). The effects of these environmental factors on plant pathogens are relatively intuitive to plant pathologists. However, predicting their effect on plant diseases per se requires greater knowledge of how these factors affect the physiology of host plant and, consequently, the host-pathogen interaction.

Effect of Elevated CO_2

It is suggested that elevated CO_2 would increase canopy size and density, resulting in a greater biomass (Manning and von Tiedemann, 1995). Combined with increased canopy humidity, it is likely to promote vector borne viruses such as Begomoviruses, Cucumoviruses, Ilarviruses, Potyviruses, and Tospoviruses.

In general, insect herbivores fed on foliage from plants grown at elevated CO_2 develop more slowly, consume more tissue, and have greater mortality (Watt *et al.*, 1995). Laboratory feeding trials on leaf-chewing insects have consistently found increased consumption rates on foliage grown at elevated CO_2 (Fajer *et al.*, 1989). The increases in consumption, on an average of 40% and to a maximum of 80% are generally considered an attempt to compensate for lower nitrogen concentrations in plant tissues grown at high CO_2 because nitrogen is single most limiting resource for insect herbivores. Similar observations were made in case of sucking insects as indicated by more amount of leaf area damaged by *Frankliniella occidentalis* at elevated CO_2 (Hughes and Bazzaz,1997).

Effect of Rise in Temperature

Studies of climate change on organisms suggest that direct effects of temperature are likely to be larger and more important than any other factor. The main effect of temperature in temperate regions is to influence winter survival of vectors. Natural spread of vectors, pests and diseases is likely to be accelerated towards the north, as former climate barriers are no longer effective. This will result in more severe outbreaks of plant disease vectors like aphids, white flies, thrips or beetles, an extension of the period of disease infection further into the growing season and also introduction and establishment of new vector species. More vectors survive from one vegetation period to next leading to earlier and faster development of transmitted disease (Boland *et al.*, 2004, Garrett *et al.*, 2006). Particularly aphids are expected to react strongly to environmental changes because of their short generation time, low developmental threshold temperatures (Harrington *et al.*, 2007) and ability to survive mild winters without winter forms. An increase in numbers of insect vectors will inevitably lead to a higher risk for viral infection of plants.

Temperature is most critical climatic variable for insects. Low temperature extremes can significantly increase mortality, while higher winter temperatures have the potential to increase the survival of many overwintering species, and may allow changes in species distribution. Higher temperatures in spring and summer could result in faster development rates and, therefore, more rapid increases of pest populations. Such effects would be particularly important in spring, when current temperatures are a limiting factor for development and distribution (Porter *et al.*, 1991).

Exceeding critical climatic thresholds affect phonological and developmental stages of insect vectors, including mortality, fecundity, oviposition and generation numbers (Kiritani, 2006). Arthropod vectors play an essential role in dissemination of viruses; more than 70% of viruses infecting plants are transmitted from one host to another by arthropod vectors (Power, 2000).

Impact on Plant Virus Pathosystems

Most research on influence of climate change on plant diseases has been concentrated on the effects of a single atmospheric variable on the host, pathogen, or the interaction of the two under controlled conditions (Coakley, 1988; Coakley and Scherm, 1996; Manning and von Tiedemann, 1995). Interactions are clearly more complex, where multiple climatological and biological factors vary simultaneously in a dynamic environment. Climate change has the potential to modify host physiology and resistance and to alter stages and rates of development of the pathogen. The most likely impacts would be shifts in geographical distribution of host and pathogen, changes in the physiology of host-pathogen interactions, and crop loss. Another important impact may be through changes in the efficacy of control strategies.

Pest outbreaks are more likely to occur with stressed plants in response to environmental extremes. The reduced resistance in plant hosts can lead to increased food and breeding site availability, particularly in tree species increasing development rates of insect pest populations.

Geographical Distribution of Host, Vector and Virus

New virus disease complexes may arise and some diseases may cease to be economically important if warming causes a poleward shift of agroclimatic zones and host plants migrate into new regions. Viruses and vectors would follow the migrating hosts and may infect remnant vegetation of natural plant communities not previously exposed to the often more aggressive strains from agricultural crops (Coakley *et al.*, 1999). The mechanism of pathogen dispersal, suitability of environment for dispersal, survival between seasons, and any change in host physiology and ecology in new environment will largely determine how quickly pathogens become established in a new region.

If the frost line moves north in the Northern Hemisphere, higher winter temperatures could be accompanied by increased survival of insects. For virus-vector aphids, this could lead to higher incidence of virus diseases, especially in those regions where the timing of virus arrival is linked to winter survival and spring flight of aphids. Barley yellow dwarf potyvirus (BYDV) is an example of a virus that causes more severe disease following mild winters. Since BYDV is exclusively vectored by aphids, increased survival of pathogen reservoirs could greatly increase the economic losses caused by infection. Similar increases in viruses of potato and sugarbeet have also been observed following warmer winters (Carter and Harrington, 1991).

Impact of Change in Climate and Monsoon System

Tropical climates are in general considered as wet and warm, but a great diversity of weather and local climates exists. Indian agriculture is largely dependent on the south-west monsoon system. Monsoons have seasonal trends, with typical season winds and rainfalls occurring from June–September. Unstable and erratic monsoon evident from changing patterns of rainfall and increased frequency of extreme events of flooding and drought are most likely to influence vector population accentuating the outbreak of viral diseases.

Physiology of Host-Pathogen Interactions

Temperature changes will affect crop productivity and plant distribution, interacting with the epidemiology and spatial distribution of their viruses and vectors, either at local or global scales. At the same time, the temperature increase and the demographic growth will affect the spread of viruses and vectors, through mechanisms related to the exploitation of new land, commerce and commodities exchange.

Increases in temperature can modify host physiology and resistance. Considerable information is available on heat-induced susceptibility and temperature-sensitive genes. The effects of elevated temperature on plants will tend to vary greatly throughout the year. During colder parts of the year, warming may relieve plant stress, whereas during hotter parts of the year it may increase stress. When high temperature stress is exacerbated, plant responses may be similar to those induced by water stress, with symptoms including wilting, leaf burn, leaf folding, and abscission, and physiological responses including changes in RNA metabolism and protein synthesis, enzymes, isoenzymes, and plant growth hormones (Christiansen and Lewis, 1982). These changes will certainly affect susceptibility to pathogens, though the wide range of changes makes interaction effects difficult to predict.

Impact on Emerging Virus Diseases

Many infectious diseases have emerged as threats to cultivated and wild plants. Some of the important emerging plant virus diseases are threatening agriculture and biodiversity conservation. Major taxonomic groups of pathogens causing emerging plant diseases show that viruses, fungi and bacteria are major pathogens causing plant EIDs. Viruses cause just under half (47%) of the reported plant EIDs, similar to that for human (44%) diseases (Taylor *et al.*, 2001). Rest of the plant EIDs are accounted by fungi (30%), bacteria (16%), phytoplasma (4%), nematode (1%) and unknown etiology (2%) for plants (Anderson *et al.*, 2004). Analysis of factors cited as the cause of disease emergence for important viruses shows that, although introduction is most important driver (71%), followed by changes in vector populations (16%). The other factors such as recombination (5%), changes in weather (5%) and changes in farming techniques (3%) are responsible for disease emergence.

Global climate change is predicted to change the distribution and abundance of arthropod vectors and to increase the frequency of unusual weather events, one of the major drivers of emerging plant viruses. Although examples of climate change driving the emergence of plant viruses have not yet been clearly identified, projected levels of climate change are likely to increase disease emergence in plants.

Emergent plant viruses can be placed into two broad categories: entire groups of viruses (e.g., genera or families) or individual viruses (Table 20.1). Some examples of plant virus groups that are emergent, on a global level, include whitefly-transmitted begomoviruses (genus *Begomoviru,* family *Geminivridae),* thrips-transmitted tospoviruses (genus *Tospovirus* family *Bunyaviridoe), Tobacco streak virus (Genus Ilarvirus,* family *Bromoviridae)* and criniviruses (genus *Crinivirus,* family *Closteroviridae).* Some individual viruses that have emerged relatively recently include the potexvirus namely *Pepino mosaic virus* (PepMV) infecting tomato. Another example is emergence of a novel whitefly-transmitted potyvirus infecting cucurbits in Florida (Adkins *et al.*, 2007).

Table 20.1. Examples of emerging viruses in horticultural crops

Virus/virus group	*Crops*	*Long distance movement*	*Changes in insect vector*	*Appearance of new species*	*Recombination/ reassertment*	*Novel virus*
Whitefly transmitted begomoviruses	Cassava, Potato, Vegetables	Yes	Yes	Yes	Yes	
Criniviruses	Tomato, Cucurbits	Yes	Yes	Yes		
Tospoviruses	Ornamental, Vegetables	Yes	Yes	Yes		
Tobacco streak virus (Ilarvirus)	Okra, Cucurbits, Ornamental	Yes	Yes	Yes		
Pepino mosaic virus (potexvirus)	Tomato				Yes	
Tomato torrado virus	Tomato					Yes

Impact on Insect Vectors

The impact of climate change on insects is very complex. Climate change could alter patterns of disturbance from vector insects through direct effects on their development and survival, adaptation capability, availability of host plants and physiological changes in host defenses, and indirect effects from changes in abundance of natural enemies, mutualists, and competitors (Porter *et al.*, 1991; Patterson *et al.*, 1999; Ayres and Lombardero 2000). Despite the caveats in climate change studies, there is already evidence that changes in insect ecology are occurring, which are consistent with predictions of climate change impacts (Harrington *et al.*, 2001).

Climate change is potentially the most important factor affecting natural and agricultural ecosystems. Its impact on the performance of herbivores is central to ameliorative or adaptive strategies. Climate change constitutes three dominant variables, the elevated CO_2, altered rainfall patterns, and temperature. The effects of these on herbivores can be through a direct influence of stress, or indirectly through effects on host plant. Temperature affects herbivorous insects directly (Bale, 2002), and it is argued that these effects will dominate the response of insects to climate change, for example, distribution of many insects is moving polewards, partly because of increased winter survival (Bale *et al.*, 2002).

The introduction of new vector species and changes in vector overwintering and oversummering (Fabre *et al.*, 2005; Mondor *et al.*, 2005; Newman 2004, 2005) and other effects of change on insects may have important influence on pathogen survival, movement, and reproduction. For example, introduction of glassy-winged sharp shooter has led to increased patterns of infection of grape plants in winter, greatly altering infection rates (Almeida *et al.*, 2005). Bale *et al.* (2002) concluded that temperature was the dominant climate factor in terms of direct effects through effects on overwintering and potentially important combination of photoperiod and temperature. There was little evidence of direct effects of CO_2 or UVB and that precipitation effects had not been studied sufficiently to draw a general conclusion. Other factors such as acid rain may also influence disease (Garrett *et al.*, 2006). Interactions between pathogens may also shift with climate change.

The importance of vector diversity for pathogen epidemiology and disease dynamics is not well-understood. Many disease agents, particularly viruses, are strongly dependent on the vectors that transmit them among hosts. Analysis of a large database of 1673 total plant viruses, including 910 unique virus species, indicated that plant viruses have much more specialized relations with their vectors than with their hosts (x^2=40.44, P<0.0001). In general, plant viruses have a low diversity of vectors and a high diversity of hosts, and the largest proportion of viruses fall into the category of host generalist/vector specialist.

A majority of vector-transmitted viruses (58.4%) have a single known vector species, whereas only 9.9% of such viruses have a single known host plant. In contrast, the proportion of host specialists among viruses without vectors (18.0%) is significantly higher. While many viruses have a very narrow range of vectors but a large host range, viruses with many vectors never have a narrow host range; no viruses fall into the category of host specialist/vector generalist. Vector specificity varies significantly according to type of vector (insects, fungi, nematodes and mites), mode of transmission (circulative or non-circulative), and virus family. Overall, these results suggest that plant disease dynamics may be more constrained by the host range of vectors than by the host range of virus. The rate-limiting step for virus spread may be attaining access to new hosts, rather than overcoming the defenses of these hosts. Thus, vector diversity is key to understanding plant virus epidemiology.

Aphids: Aphids are among most serious agricultural insect pests. They cause major economic losses in several crops worldwide, directly because of their feeding and indirectly by inflicting plant impairments (e.g., viruses, phytoxemias). However, their population threshold level as virus vectors is much lower than as direct feeding pests. Aphids are most common vectors of plant viruses, and aphid-borne nonpersistently transmitted viral diseases are of greatest economic importance in several annual cropping systems (Tomlinson, 1987). Aphids transmit about 50% of approximately 600 viruses with invertebrate vectors and most of the roughly 290 known aphid borne viruses are non-persistent (NPV) (Hull, 2002).

The economically important aphid borne viruses occurring on horticultural crops prevalent in India are Banana bract mosaic virus, banana bunchy top virus, bean common mosaic virus, Bean yellow mosaic virus, cardamom mosaic virus, chilli veinal mottle virus, cowpea aphid borne virus, cucumber mosaic virus, Dasheen mosaic virus, papaya ring spot virus –P, papaya ring spot virus –W, potato leaf roll virus, Potato virus Y, sweet potato feathery mottle virus, watermelon mosaic virus, and Zucchini yellow mosaic virus. The pest status of these small insects is related to several features of their biology that enable them to locate and exploit their host plants. The evolution of parthenogenesis, enabling reproduction without males for part or all of the life cycle, conferred a clear reproductive advantage by doubling the intrinsic rate of population increase (*rm*) relative to sexually reproducing individuals. Parthenogenesis has also conferred the ability to start development at ovulation, shortening the time between birth and reproductive maturity (Powell *et al.*, 2006).

Aphids, because of their short generation time and low developmental threshold temperatures, are an insect group expected to respond particularly strongly to environmental changes (Harrington *et al.*, 2007). Of the approximately 4,400 known species, around 250 feed on agricultural and horticultural crops (Blackman and Eastop, 2000) where they can cause significant economic damage through removal of phloem sap and by other means including, together with transient species in search of host plants and transmission of plant viruses. Continuously parthenogenetic aphids may thus achieve 18 generations a year in the United Kingdom giving the potential for a sensational volume (Harrington, 1994) or weight (Karley *et al.*, 2004) of aphids. With a warming of 2 °C an extra five generations a year might be expected (Yamamura and Kiritani, 1998). Although, a range of abiotic and biotic constraints prevents aphids from attaining anywhere near their full population growth potential, it is clear that climatic conditions, especially temperature, are immensely influential in determining their dynamics, and climatic changes are likely to affect strongly their pest and vector status.

Insects such as aphids are expected to have increased survival with milder winter temperatures, and higher spring and summer temperatures will increase their development and reproductive rates and lead to more severe disease (Alonso-Prados *et al.*, 2003). Milder winters are also expected to increase survival of alternate weed hosts of viruses. Increases in frequency and intensity of summer storms with high winds, rain, and hail will increase wounding of plants and result in increased transmission of viruses by mechanical means. Therefore, with predicted changes in climate, viral diseases of plants are expected to increase in importance. Potentially of greater importance will be the effects of diseases caused by new, introduced viruses that, because of the changed climate, will be able to persist. A warmer climate might also allow viruses that are present in greenhouses, such as *Pepino mosaic virus* (PepMV), to establish infection in the field (French *et al.*, 2001).

Thrips: Thrips transmit plant viruses in the tospovirus, Ilarvirus, carmovirus, sobemovirus and Machlomovirus genera. Tospoviruses are the cause of a number of significant emerging diseases, such as capsicum chlorosis and scape blight of onion. They infect thrips as well as plant hosts and the relationship between pathogen and vector is intimate. Once infected at the larval stage, adult thrips usually transmit tospoviruses for life. Thrips-transmitted viruses can cause significant diseases of many crop plants and their impact worldwide is immense. Onion thrips are also an important vector of several plant viruses such as tomato spotted wilt virus (Kritzman *et al.*, 2002; Jenser *et al.*, 2003), tobacco streak virus (Sdoodee and Teakle, 1987). In India, *T. tabaci* and *T. palmi* are important in terms of pest and vectors of tospoviruses. Increasingly, *T. palmi* has become an important pest around the world as it has spread within tropical regions of Africa, Australia, South America, Hawaii and the Caribbean, and to sub-tropical Florida and Japan.

The adaptive diversity of thrips has enabled successful exploitation of diverse niches resulting in their establishment in a variety of plants. The increasing evidence of thrips infestation in various cropping systems, their ability to migrate from weed reservoirs to crops and *vice versa*, their intercrop movement and related adaptive strategies are basic aspects of thrips bionomics. Their abundance and patterns of distribution on various host-plants involving both crops and weeds are important factors in their population dynamics. Successful interactions of thrips with their host-plants appear to depend on a complex set of environmental, visual, tactile and chemical factors that appreciably influence their behaviour and physiology.

Although temperature is the main factor affecting onion thrips development, it is an open question of whether the thermal constants are truly constant in different climatic conditions. Different values for thermal constants of onion thrips, reported in the literature (Edelson and Magaro 1988; Murai 2000; Stacey and Fellowes 2002), could be partially related to the adaptation capacity of onion thrips. Nevertheless, we can expect an increase in number of generations of onion thrips per season because of the big predicted increase in cumulative degreedays and the prolongation of the period with favorable developmental conditions. More generations could lead to larger populations and consequently more damage on host plants. In Slovenia, especially the areas already grapple with the problem of onion thrips. The onion thrips also could become a serious threat in areas where it is not currently a problem. Bergant *et al.* (2005) quantified the impact of future climate change on the developmental dynamics of onion thrips in Lovenia. Monthly averaged results of simulations of future climate from four different general circulation models (GCMs) were projected to local scale by empirical downscaling.

The reproduction and development of onion thrips are strongly related to environmental conditions, especially temperature (Edelson and Magaro, 1988).

Whiteflies: The whitefly, *Bemisia tabaci* (Gennadius) (Homoptera: Aleyrodidae), is an agricultural pest in tropical and subtropical regions and has colonized plants in greenhouses of semi-temperate countries. *Bemisia tabaci* has increased in importance globally since the 1990s, as a serious pest of vegetable, fibre, and root crops (Duffus, 1987; Polston and Anderson, 1997; Varma and Malathi, 2003), because of polyphagous nature of some biotypes and diverse ways that it damages crops. These are direct feeding, inducing phytotoxic disorders and, most significantly, by vectoring numerous plant viruses (Muniyappa, 1980; Duffus, 1987; Maruthi *et al.,* 2002; Jones, 2003; Varma and Malathi, 2003). The majority of these viruses belong to the genus *Begomovirus* of family Geminiviridae (Varma and Malathi, 2003), which causes some of the world's most economically important plant-virus diseases. The increased importance of begomoviruses as new and emerging plant viral pathogens is directly related to adaptive capacity of *B. tabaci* and its ability to exploit agricultural systems, which increasingly impose upon their subtropical habitats. The *B. tabaci* has long confounded systematists, undermined agricultural production, and proven difficult to control partly because of its propensity to develop insecticide resistance.

Areas where *B. tabaci* has a significant economic impact are found in equatorial Africa, where it is associated with the intense spread of African cassava mosaic virus (ACMV) and in Pakistan and India where it is involved in the transmission of economically important begomoviruses in vegetable crops (Simón *et al.*, 2003; Rekha *et al.*, 2005). It is also a significant pest in vegetable crops of the southern USA and the Caribbean Basin (Brown *et al.*, 2001).

The B biotype was first recorded in Kolar district of Karnataka, south India, during summer growing season (March–June) of 1999 (Banks *et al.*, 2001). The economic importance of

B-biotype became immediately evident when numbers of whiteflies on tomato, cabbage, and other vegetable crops, which previously were seldom infested by local whiteflies, increased 1,000-fold. The most significant damage was due to an epidemic of tomato leaf curl virus disease, which has since continued to affect tomato production seriously in India (Banks *et al.*, 2001). The data obtained on the relatively B biotype is now present in most parts of south India, which may suggest that it has spread from the original point of introduction in Kolar district. Since 1999, the B-biotype has spread about 600 km (Mysore to Raichur) in north-south and about 250 km (Shimoga to Chittor) in the east-west directions (Rekha *et al.*, 2005). The researchers report that invasive whiteflies are successful at least in part because they breed more when they come into contact with native whiteflies, and they also suppress the native whiteflies' reproduction rates.

IMPACT ON DISEASE MANAGEMENT

Inadequate knowledge about impacts of climate change on plant pathosystems, makes it difficult to predict implications for disease management with any certainty. It is prudent to assume, however, that effects would occur chiefly through influences on host resistance or chemical and biological control agents. Particular attention is needed to identify cases where the efficacy of disease management may be reduced under climate change.

Host resistance: Cultivar resistance to pathogens may become more effective because of increased static and dynamic defenses from changes in physiology, nutritional status, and water availability. Durability of resistance may be threatened, however, if number of infection cycles within a growing season increases because of one or more of the following factors: increased fecundity, more pathogen generations per season, or a more suitable microclimate for disease development. Temperature may have important repercussions on the effectiveness of resistance genes, though it may generally be challenging to discriminate between temperature effects on host resistance genes versus effects on pathogen virulence. In India, resistance to Okra yellow vein mosaic virus tends to break when okra is grown at higher temperatures. Thrall and Burdon (2003) have found important variation in both host resistance and pathogen virulence among natural populations. If climate change increases or decreases environmental conduciveness, the shift in selection pressure on the host populations could result in shifts in the diversity of resistance genes present.

Plant protection: Climate change could affect the efficacy of crop protection chemicals in one of two ways. First, changes in temperature and precipitation may alter the dynamics of pesticide residues on the crop foliage. Globally, climate change models project an increase in the frequency of intense rainfall events, which could result in increased fungicide wash-off and reduced control. The interactions of precipitation frequency, intensity, and pesticide dynamics are complex, and for certain pesticides precipitation following application may result in enhanced disease control because of a redistribution of the active ingredient on the foliage.

Impact on models: Plant viruses and vectors are extremely dependent on climatic and environmental conditions; several passes of a virus and vector life cycle strictly depend on the combination of two or more environmental variables. Optimal temperature and moisture are key elements for modeling and forecasting the extent of virus/vector epidemics, or to develop preventive crop protection or management strategies. Most of what has been said about plant disease in relation to climate change is based on qualitative, rule-based reasoning. This approach seems attractive because of the substantial body of knowledge already available regarding the environmental requirements of plants and their pathogens (Chakraborty *et al.*, 2000). However, quantitative prediction will depend on complex interactions of atmospheric, climatic, and biological factors with technological and socioeconomic changes that are exceedingly difficult to predict (Chakraborty *et al.*, 2000). It would appear, therefore, that in all but the simplest cases these interactions are not amenable to qualitative analyses. Hence, quantitative (modeling) approaches, which allow one to investigate multiple scenarios and interactions simultaneously, will become more important for impact assessment.

ABERRANT WEATHER AND EXTREME EVENTS

- In Andhra Pradesh, after floods during 2006 heavy incidence of chilli veinal mottle virus and cucumber mosaic virus was observed causing severe yield loss of chilli in Khambham and Guntur districts.
- In India, tobacco streak virus occurred after severe drought during 1997 on sunflower crop in Karnataka. Soon this virus shifted to groundnut in Andhra Pradesh, causing severe yield loss during 2002. Later this virus started appearing epidemic form on vegetable crops causing epidemic in gherkin and okra in Karnataka (Krishnareddy *et al.,* 2003a, 2003b).
- The B-biotype of whiteflies was first recorded in the Kolar district of Karnataka, south India, during summer growing season (March–June) of 1999. Soon the numbers of whiteflies increased 1,000-fold on tomato, cabbage, and other vegetable crops seldom infested by local whiteflies. An epidemic of tomato leaf curl virus disease followed, which has since continued to affect tomato production seriously in India. Since 1999, the B biotype has spread about 600 km (Mysore to Raichur) in north-south and about 250 km (Shimoga to Chittor) in the east-west directions (Rekha *et al.*, 2005).

Adaptation Measures for Climate Change and Global Warming

- Changing crop rotation and choosing resistant varieties.
- Monitoring the traits of "new" as well as "old" plant diseases. Such monitoring is a prerequisite for plant breeders developing new varieties with the right traits for resistance.
- Providing ready access to farmers on updated information on disease resistance in the varieties that they intend to grow.
- Breeding disease, pest and drought resistant varieties.
- Increasing the effectiveness of disease management strategies.
- Careful tracking of geographical distribution of plant virus diseases and their vectors.
- Developing early disease fore warning models.
- Use of biological control measures.
- Pest risk analysis to account for climate change effects.
- Phytosanitary regulations to prevent or limit the introduction of risky plant viruses.

Mitigation Measures to Reduce Emission of CO_2/GHG Carbon Sequestration Potential of Crops

Emphasis on biological control of diseases and agronomic management to avoid multiple chemical sprays involving tractor mounted implements.

FUTURE STRATEGIES FOR RESEARCH

- Impact assessment of climate change on important and emerging virus diseases and vectors through regional surveys.
- Develop disease appearance models for important virus and vectors.
- Intensification of virus, vector and reservoir surveillance.
- Systematic observations on aberrant weather conditions in relation to virus and vector outbreak.
- Identification of key environmental factors that influence the geographical distribution of arthropod vectors and their relationships with the viruses they transmit.
- At the molecular level, develop genomics tools that will enable us to examine the impact of elevated temperatures on virus infection of plants and virus transmission by arthropod vectors
- Studies on climate change and effects on introduction and spread of plant viruses and their vectors.
- Improve understanding of gene expression in plants and pathogens in response to climatic factors
- Integrated multidisciplinary national and international networks for data collection and synthesis.

CONCLUSION

Plant virus diseases cause significant loss to India's agricultural and horticultural crops. The economic impact of disease results from losses in productivity, the cost of disease management and economic penalty as a result of growing less profitable alternative crops. Most of the plant virus diseases are spread by insect vectors mainly aphids, thrips and whiteflies. Weather and climate strongly influences plant virus diseases and vectors. The climate change particularly global warming is expected to influence vector establishment, spread and reproduction potential as well as transmission of viruses. It will result in more severe outbreaks of plant-disease vectors like aphids, white flies, thrips and beetles, an extension of the period of disease infection further into the growing season and also introduction and establishment of new vector species. Particularly aphids are expected to react strongly to environmental changes because of their short generation time, low developmental threshold temperatures and ability to survive mild winters without winter forms. Aphids are expected to have increased survival with milder winter temperatures, and higher spring and summer temperatures will increase their development and reproductive rates and lead to more severe disease. Climate change could alter patterns of disturbance from vector insects through direct effects on their development and survival, adaptation capability, availability of host plants and physiological changes in host defenses, and indirect effects from changes in the abundance of natural enemies, mutualists, and competitors. An increase in numbers of insect vectors will inevitably lead to a higher risk for viral infection of plants.

Climate change particularly warming could also alter stages and rates of development of the pathogen, modify host resistance, and result in changes in physiology of host-pathogen interactions. With warming, some virus diseases may be able to spread faster or at higher elevations than under current climate conditions. The host resistance to viruses may breakdown more rapidly owing to accelerated virus evolution, a result of the shorter life cycles of insect vectors, recombination and mutation of virus strains.

Prediction of disease outbreaks and vectors will be more difficult in periods of rapidly changing climate and unstable weather. Globally, climate change models project an increase in the frequency of intense rainfall events, which could result in increased fungicide wash-off and reduced control. The interactions of precipitation frequency, intensity, and pesticide dynamics are complex, and for certain pesticides precipitation following application may result in enhanced disease control because of a redistribution of the active ingredient on the foliage. Hence, quantitative (modeling) approaches, which allow one to investigate multiple scenarios and interactions simultaneously, will become more important for impact assessment and evolving control strategies.

References

Adkins, S., Webb, S.E., Achor, D., Roberts, P.D., Baker, C.A. 2007. Identification and characterization of a novel whitefly-transmitted member of the family *Potyviridae* isolated from cucurbits in Florida. *Phytopathology* **97**:145-54.

Almeida, R.P.P., Wistrom, C., Hill, B.L., Hashim, J., Purcell, A.H. 2005. Vector transmission of *Xylella fastidiosa* to dormant grape. *Plant Dis.* **89**:419-24.

Alonso-Prados, J.L., Luis-Arteaga, M., Alvarez, J.M., Moriones, E., Batlle, A., Lavina, A., Garcia-Arenal, F. and Fraile, A. 2003. Epidemics of aphid-transmitted viruses in melon crops in Spain. *Eur. J. Plant Pathol.* **109**:129-38.

Ayres, M.P. and Lombardero, M.J. 2000. Assessing the consequences of global change for forest disturbance from herbivores and pathogens. *Sci. Total. Environ.* **262**:263-86.

Bale, J.S. 2002. Insects and low temperatures: from molecular biology to distributions and abundance. Philosophical Transactions of the Royal Society of London Series B – *Biological Sciences* **357**:849-61.

Bale, J.S., Masters, G.J, Hodkinson, I.D., Awmack, C., Bezemer, T.M. 2002. Herbivory in global climate change research: direct effects of rising temperature on insect herbivores. *Global Change Biology* **8**:1-16.

Banks, G.K., Colvin, J., Chowda Reddy, R.V., Maruthi, M.N., Muniyappa, V., Venkatesh, H.M., Kiran Kumar, M., Padmaja, A.S., Beitia, F.J. and Seal, S.E. 2001. First report of the Bemisia tabaci B biotype in India and an associated tomato leaf curl virus disease epidemic. *Plant Disease* **85**:231.

Bergant, K., Tradan, S., Znidarcic, D., Crepinsek Z and Kajfez-Bogataj L. 2005. Impact of Climate Change on Developmental Dynamics of *Thrips tabaci* (Thysanoptera: Thripidae): Can it be quantified? *Environ. Entomol.* **34**:755-66.

Blackman, R.L. and Eastop, V.F. 2000. *Aphids on the World's Crops.* Wiley, Chichester.

Boland, G.J., Melzer, M.S., Hopkin, A., Higgins, V. and Nassuth, A. 2004. Climate change and plant diseases in Ontario. *Can. J. Plant Pathol.* **26**:335-50.

Brown, J.K., Frohlich, D.R. and Rosell, R.C. 1995. The sweet potato or silver-leaf whiteflies: biotypes of *Bemisia tabaci* or a species complex? *Annual Review of Entomology* **40**:511-34.

Bruinsma, J. 2008. Climate Change and agriculture: Physical and human dimensions. *In: World Agriculture: Toward 2010/2030*; An FAO perspective, FAO, Eart scale Publication Pvt. Ltd., London, p. 357-72.

Carter, N., Harrington, R. 1991. Factors influencing aphid population dynamics and behavior and the consequences for virus spread. *In: Advances in Disease Vector Research.* Harris, K.F., (Ed.). **7**:19-51. Springer, New York.

Chakraborty, S., Tiedemann, A.V., and Teng, P.S. 2000. Climate change: potential impact on plant diseases. *Environ. Pollution* **108**:317-26.

Coakley, S.M.1988. Variation in climate and prediction of disease in plants. *Annu. Rev. Phytopath.* **26**:163-81.

Coakley, S.M. and Scherm, H. 1996. Plant disease in a changing global environment. *Asp. Appl. Biol.* **45:**227-38.

Coakley, S.M., Scherm, H. and Chakraborty, S. 1999. Climate change and plant disease management. *Annu. Rev. Phytopath.* **37**:399-26.

Duffus, J.K.1987. Whitefly transmission of plant viruses. *Current Topics in Vector Research* **4**:73-91.

Edelson, J.V. and Magaro, J.J. 1988. Development of onion thrips, Thrips tabaci Lindeman, as a function of temperature. *Southwest Entomol.* **13**:171-76.

Fabre, F., Plantegenest, M., Mieuzet, L., Dedryver, C.A., Leterrier, J.L. and Jacquot, E. 2005. Effects of climate and land use on the occurrence of viruliferous aphids and the epidemiology of barley yellow dwarf disease. *Agric. Ecosyst. Environ.* **106**:49-55.

Fajer, E.D., Bowers, M.D. and Bazzaz, F.A. 1989. The effects of enriched carbon dioxide atmospheres on plant-insect herbivore interactions. *Science* **243**:1198-200.

French, C.J., Bouthillier, M., Bernardy, M., Ferguson, G., Sabourin, M., Johnson, R.C., Masters, C., Godkin, S. and Mumford, R. 2001. First report of *Pepino mosaic virus* in Canada and the United States. *Plant Dis.* **85**:1121.

Garrett, K.A., Dendy, S.P., Frank, E.E., Rouse, M.N. and Travers, S.E. 2006. Climate Change Effects on Plant Disease:Genomes to Ecosystems. *Annu. Rev. Phytopathol.* **44**:489-509.

Harrington, R. 1994. Aphid layer. *Antenna* **18:**50-51.

Harrington, R., Clark, S.J., Welham, S.J., Verrier, P.J., Denholm, C.H., Hulle, M., Maurice, D., Rounsevell, M.D. and Cocu, C. 2007. Environmental change and the phenology of European aphids. *Global Change Biology* **13**:1550-64.

Harrington, R., Fleming, R.A. and Woiwod, I.P. 2001. Climate change impacts on insect management and conservation in temperate regions: can they be predicted? *Agric. for. Entomol.* **3**:233-40.

Harvell, C.D., Mitchell, C.E., Ward, J.R., Altizer, S. and Dobson, A.P. 2002. Climate warming and disease risks for terrestrial and marine biota. *Science* **296:**2158-62.

Hughes, L. and Bazzaz, F.A. 1997. Effect of elevated CO_2 on interactions between the western flower thrips, *Frankliniella occidentalis* (Thysanoptera: Thripidae) and the common milkweed, Asclepias syriaca. *Oecologia* **109**:286-90.

Hull, R. 2002. Matthews' Plant Virology. Academic Press, San Diego, California, pp.1001.

IPCC. 2001. *Climate Change 2001: The Scientific Basis. Contrib. Work. GroupI Third Assess. Rep. Intergov. Panel Clim. Change*, Houghton, J.T., Ding, Y., Griggs, D.J., Noguer, M., van der Linden, P.J., *et al.* (Eds). Cambridge, UK/ New York: Cambridge Univ. Press.

Jenser, G., Ga´borja´nyi R, Sze´nasi A, Alma´asi A and Grasselli M. 2003. SigniÞcance of hibernated *Thrips tabaci* (Thysan., Thripidae) adults in the epidemic of tomato spotted wilt virus. *J. Appl. Entomol.* **127**:7-11.

Karley, A.J., Parker, W.E., Pitchford, J.W. and Douglas, A.E. 2004. The midseason crash in aphid populations: why and how does it occur? *Ecological Entomology*, **29:**383-88.

Kiritani, K. 2006. Predicting impact of global warming on population dynamics and distribution of arthropods in *Japan. Population Ecology* **48**:5-12.

Krishnareddy, M., Devaraj, Lakshmi, R., Jalai, S. and Samuel, D.K. 2003a. Outbreak of *Tobacco streak virus* causing necrosis of cucumber (*Cucumis sativus*) and Gherkin (*Cucumis anguria*) in India. *Plant Dis.* **87**:1264.

Krishnareddy, M., Jalai, S., Samuel, D.K. 2003b. Fruit distortion mosaic disease of Okra in India. *Plant Dis.* **87**:1395.

Kritzman, A., Gera, A., Raccah, B., van Lent J.W.M., and Peters, D. 2002. The route of tomato spotted wild virus inside the thrips body in relation to transmission efficiency. *Arch. Virol.* **147**:2143-56.

Manning, W.J. and Von Tiedemann, A. 1995. Climate change: potential effects of increased atmospheric carbon dioxide (CO_2), ozone (O_3), and ultraviolet-B (UV-B) radiation on plant diseases. *Environ. Pollut.* **88**:219-45.

Maruthi, M.N., Colvin, J., Seal, S., Gibson, G. and Cooper, J. 2002. Co-adaptation between cassava mosaic geminiviruses and their local vector populations. *Virus Research* **86**:71-85.

Mondor, E.B., Tremblay, M.N., Awmack. C.S. and Lindroth, R.L. 2005. Altered genotypic and phenotypic frequencies of aphid populations under enriched CO2 and O3 atmospheres. *Glob. Change Biol.* **11**:1990-96.

Muniyappa, V. 1980. Whiteflies. *In:Vectors of Plant Pathogens.* Harris, K.F. and Maramorosch, K(Eds). pp. 39-85. Academic Press, New York, USA.

Murai, T. 2000. Effect of temperature on development and reproduction of onion thrips, *Thrips tabaci* Lindeman (Thysanoptera: Thripidae), on pollen and honey solution. *Appl. Entomol. Zool.* **35**: 499-504.

Newman, J.A. 2004. Climate change and cereal aphids: the relative effects of increasing CO2 and temperature on aphid pollution dynamics. *Glob. Change Biol.* **10**:5-15.

Newman, J.A. 2005. Climate change and the fate of cereal aphids in Southern Britain. *Glob. Change Biol.* **11**:940-44.

Patterson, D.T., Westbrook, J.K., Joyce, R.J.V., Lingren, P.D. and Rogasik, J. 1999. Weeds, insects, and diseases. *Clim. Change.* **43**:711-27.

Polston, J.E. and Anderson, P.K. 1997. The emergence of whitefly-transmitted geminiviruses in tomato in the Western Hemisphere. *Plant Dis.* **81**:1358-69.

Porter, J.H., Parry, M.L. and Carter, T.R. 1991. The potential effects of climatic change on agricultural insect pests. *Agric. For. Meteorol.* **57**:221-40.

Powell, G., Tosh, C.R. and Hardie, J. 2006. Host plant selection by aphids: Behavioral, evolutionary and applied perspectives. *Annu. Rev. Entomol.*, **51**:309-30.

Power, A.G. 2000. Insect transmission of plant viruses: a constraint on virus variability. Curr. *Op. Plant Biol.* **3**:336-40.

Rekha, A.R., Maruthi, M.N., Muniyappa, V. and Colvin, J. 2005. Occurrence of three genotypic clusters of *Bemisia tabaci* and the rapid spread of the B biotype in south India. *Entomologia Experimentalis et Applicata* **117**:221-33.

Sandermann, H. Jr. 2000. Ozone/biotic disease interactions: molecular biomarkers as a new experimental tool. *Environ. Pollut.* **108**:327-32.

Sdoodee, R. and Teakle, D.S. 1987. Transmission of tobacco streak virus by *Thrips tabaci:* a new method of plant virus transmission. *Plant Pathol.* **36**:377-80.

Simon, B., Cenis, J.L., Beitia, F., Khalid, S., Moreno, I.M., Fraile, A. and Arenal, F.G. 2003. Genetic structure of field populations of begomoviruses and of their vector *Bemisia tabaci* in Pakistan. *Phytopathology* **93**:1422-29.

Stacey, D.A. and Fellowes, M.D.E. 2002. Temperature and the development rates of thrips: evidence for constraint on local adaptation? *Eur. J. Entomol.* **99**:399-404.

Thrall, P.H. and Burdon, J.J. 2003. Evolution of virulence in a plant host-pathogen metapopulation. *Science* **299**:1735-37.

Tomlinson, J.A. 1987. Epidemiology and control of virus diseases of vegetables. *Ann. Appl. Biol.* **110**:661-81.

Varma, A. and Malathi, V.G. 2003. Emerging geminivirus problems. A serious threat to crop production. *Annals of Applied Biology* **142**:145-64.

Yamamura, K. and Kiritani, K. 1998. A simple method to estimate the potential increase in the number of generations under global warming in temperate zones. *Applied Entomology and Zoology* **33**:289-98.

21

Implications of Climate Change on Horticultural Crop Pests — Indian Scenario

V. Sridhar, N. K. Krishna Kumar and Abraham Verghese

THERE is a growing concern about the global warming caused by excess emission of several naturally occurring gases like carbon dioxide, methane, nitrous oxide and ozone. Intergovernmental Panel on Climate Change (IPCC) projects that global mean temperature may increase between 1.4 and 5.8 °C by 2100. It is also assumed that there will be an increase of 0.1 °C in both minimum and maximum temperatures for each degree of latitude. Thus, greater warming is expected at higher latitudes than in the tropics.

The geographic distribution, vigour, virulence, and impact of weeds, insects, and plant pathogens on agriculture will be influenced by climatic changes, accompanying the global greenhouse effect. Physiological and biochemical changes induced in host crop plants by rising CO_2 may affect feeding patterns of insect pests. Pest generation times may be reduced, enabling more rapid population increases to occur. Poleward migration may be accelerated during the crop season. The epidemiology of plant diseases also will be altered. Prediction of disease outbreaks will be more difficult in periods of rapidly changing climate and unstable weather particularly where insects are involved in the transmission of diseases.

Environmental instability and increased incidence of extreme weather may reduce the effectiveness of pesticides on targeted pests or result in more injury to non-target organisms. Biological control may be affected either negatively or positively. Periurbanisation is the order of the day. Enhanced CO_2 emission along with other pollutants coupled with high temperature will result in the complex changes in the pest systems, particularly under horticultural ecosystems. Among vegetables, sucking pests are likely to increase such as thrips, mites, and leafhoppers with the increasing temperatures. Overall, challenge to agriculture from pests probably will increase (Patterson *et al.*, 1999).

Various possible implications of global warming on insect pests is being reviewed in this chapter based on various works carried out world over to know the effect of climate change on insect pests and disease vectors with particular reference to its possible effect on pest incidence on Horticultural crops in India.

IMPACT ANALYSIS ON CROP PESTS INDIAN SCENARIO

Various climate change factors might affect the pests either positively or negatively. Various implications of climate change on pests are categorised as follows.

Direct effect of climate change on pests

- Expansion of habitat range of Pests
- Change in migrating behaviour
- Changes in over wintering success
- Changes in interaction between species.
- Effect of pest-natural enemy interactions

Indirect effect through host plants

- Effect of increased temperature
- Effect of increased CO_2
- Effect of increased pollutants (ozone and nitrous oxide)
- Effect of changes in host plant distribution

Important direct and indirect effects of climate change on pest of horticultural crops are discussed below.

Elevated CO_2 and Temperature

CO_2

Guerenstein and Hildebrand (2008) have reviewed the role and effects of environmental carbon dioxide on insect life. Free air concentration enrichment (FACE) studies showed that during the early season, soybean grown in elevated CO_2 had 57% more insect damage than those in current levels and required insecticide treatment in order to continue the experiment. Measured increases in the levels of simple sugars in soybean leaves might have stimulated the additional insect feeding.

Increased carbon to nitrogen ratios in plant tissue resulting from increased CO_2 levels may slow insect development and increase the length of life stages vulnerable to attack by parasitoids. Further, negative response of insects to elevated CO_2 is that they accumulate more carbon, with an increase of C:N ratio, so that insects are faced with nutritionally deficient host plants resulting in a N-dilution effect (Ananthakrishnan, 2007). Thus, under higher CO_2 conditions feeding by herbivores will be high to derive more amino acids. For example, *Pseudoplusia includens* feeding was 30% greater on leaves from the 650-µl/litre of CO_2 treatments than on leaves from the 350 µl/litre treatment. This study suggests that feeding by herbivores on leaves of C_3 plants may increase as the level of atmospheric carbon dioxide rises (Lincoln *et al.*, 1986). Moreover, at elevated CO_2 there is greater partitioning of assimilates to roots in crops such as carrot, radish and sugarbeet. Due to increased carbon storage in roots, losses from soil-borne pests may diminish under climate change (Coakley *et al.*, 1999).

Similarly in case of *Trichoplusia ni* elevated atmospheric carbon dioxide levels of 1000 ppm significantly increased consumption of foliage compared with ambient levels of 340 ppm The percentage leaf area of plants consumed by *T. ni* larvae was not affected by carbon dioxide concentration, suggesting that increased plant growth resulting from raised atmospheric carbon dioxide may benefit the plant proportionately more than the insect (Osbrink *et al.*, 1987).

Brevicoryne brassicae reared on *Brassica oleracea* produced significantly less offspring at elevated CO_2, whereas the opposite was found for *M. persicae* on the same host. No response was found for *M. persicae* on *Senecio vulgaris*. When populations of *B. brassicae* and *M. persicae* were followed for a longer period, no differences were observed in population sizes. Comparisons between different experimental systems showed that long-term population responses to elevated CO_2 could not be reliably predicted from detailed measurements on individual aphids (Bezemer *et al.*, 1999).

Increased Temperature

Insect physiology and host pest interaction: Insects are poikilothermic in nature, i.e., their surrounding temperatures determine their development. The increased temperature limit their range, over wintering, populations growth rate, number of generations per annum, length of growing season, crop-pest synchronization, intraspecific interaction, dispersal, migration and availability of host plants and may alter host physiology and resistance. Elevated temperatures may cause breakdown in temperature sensitive resistance. Increase in food quality due to increase in nitrogen content in plants under stress conditions caused by high temperature can result in sudden resurgence of insect pests population. Moreover, defensive system of plants is lowered under stress conditions, and they become more susceptible to pest attacks. Conversely, in some forage species, there is increased lignification at higher temperatures, which can enhance the level of host resistance to pathogens. Insects may gain or lose based on the prevailing temperatures in which they are living.

Insect population: With the climate change, under tropical conditions the winter duration will be reduced thereby affecting the hibernation of insects resulting in more multiplication of pests with more number of generations. When temperatures are above 35 °C lepidopterous pests may be reduced and smaller pests like leaf miner, thrips, mites and whiteflies may increase in numbers. However, under temperate conditions, caterpillar pests may increase with the change in climate.

Under Indian conditions, diamond back moth, *Plutella xylostella*, a major pest on cruciferous vegetables may increase up to 28–35 °C range and thereafter in areas where temperature increases further, the incidence may decrease. Similarly, leaf miner- *(Liriomyza)* incidence will increases up to 30–35 °C. Incidence of vector borne diseases transmitted by thrips, whitefly, mealy bug, red spider mite, etc., may increase with the increases in average temperatures. Development of *P. xylostella* takes 15 days at 25 °C whereas it completes its development in 14.3 days only at 27.5 °C (Yamada and Kawasaki, 1983). Thus with the increase in temperature, with global climate change, this pest may develop 2-3 more additional generations per year. Atwal (1955) investigated the factors influencing fecundity and reported that females reared at low temperatures (7–24 °C) laid more eggs than those reared at higher temperatures (28–35 °C).

Global warming may be responsible for the recent decline in abundance of *P. xylostella* and increase in *Helicoverpa armigera* and *Trichoplusia ni* in some regions. In general, global warming may work in favour of natural enemies (except for spiders) by increasing the number of generations more than in their host species. Biological control utilizing native natural enemies is expected to become a more important control tactic in future (Kiritani, 2006).

Yamaguchi *et al.* (2001) examined the effects of a rise in temperature of 2-3 °C on the phenology and occurrence of arthropod crop pests using data obtained from Kagoshima in southwestern Japan, in 1998 when the average temperature was 2 °C higher than that between 1960 and 1998. Several arthropod species (*Homona magnanima, Adoxophyes honmai, Scirtothrips dorsalis*, *Pseudaulacaspis pentagona*, aphids (mainly *Myzus persicae* and *Lipaphis erysimi*), and *Tetranychus kanzawai*) emerged earlier than in previous years because of unusually high average temperature. All species showed developmental zeros of less than 10 °C. However, *Spodoptera litura* and *Helicoverpa armigera*, whose developmental zeros were above the mean winter temperature, emerged as usual in the spring.

A possible increase in number of generations per year was not clearly shown by the trap capture records, but all species were likely to have 1-2 more generations. The total number of male *Spodoptera litura* captured by pheromone traps in 1998 was the largest in past 14 years. There was a positive correlation between the abundance of males and mean annual temperatures. In contrast, *Homona magnanima*, *A. honmai*, and aphids were less abundant in 1998. No detectable difference in abundance was observed in *Helicoverpa armigera*. It was concluded that an elevation in mean temperature of 2 °C would impact upon the phenology of insects, but the pattern of response would vary greatly among species.

In India, whitefly, *Bemisia tabaci* was a minor pest on potato till recently. Data on population build-up during the last 20 years revealed that average population of *B. tabaci* was 11 whiteflies/100 leaves during 1984, which rose to 24.24 in 2004. During this period, average ambient temperature increased by 1.07 °C. This could be because of climate warming leading to whitefly infestation in Indo Gangetic plains. Whitefly populations led to outbreak of a new viral disease known as apical leaf curl in potato. Similarly, leafhopper (*Empoasca fabae*) and aphid *Myzus persicae* populations are also on rise in the recent years.

Pest migration: With the change in temperatures, new pests (moths and butter flies) will migrate to new areas. Already introduced pests like American leaf miner, *Liriomyza* may spread to new areas northwards with the increase in temperatures. Non-indigenous pests may become established initially in protected crops, as a lot of plant material is being imported for protected cultivation, particularly, ornamentals. Gradually, these pests will establish in field crops subsequently, as the climate changes.

The increasing temperature in temperate latitudes will result in less cold stress from low minimum temperatures, longer growth seasons with more degree-days for development and more heat stress of temperate species from higher maximum temperatures. As these changes take place, day lengths remain unchanged so species with close linkages to day lengths will have their dormancy, diapause, migration or reproductive processes disrupted. In addition, they will be at risk of a loss of synchrony with other species of plants and animals with which they interact but which respond differently to temperature changes. All of these changes will contribute to shifts in geographical distributions of pests. Temperate species will push further towards the Polar Regions while tropical and sub-tropical species will encroach on the temperate regions.

In the tropics the most likely changes of biological significance will be related to changes in available moisture. With lower average rainfall scenarios, and more intense rainfall dispersed between longer dry periods, the seasonal growth of many pests and their associated biota are likely to alter. More prolonged and intensive droughts are expected. Crop pests and pathogens may be affected by changes in the crop canopies as enhanced concentration of CO_2 stimulate growth of foliage in situations where other nutrients are not limiting.

In India, there will be a change in natural distribution range shift of pests like fruit flies, *Bactrocera* sp., green stink bug, *Nezara viridula*, sweet potato weevil, *Cylas formicarius* which are presently in tropical and subtropical climates and may shift to temperature zone. Some of the pests like blossom midge, *Contarinia maculipennis* (Diptera: Cecidomyiidae) may result in expansion of host range from orchids to various crops.

It is predicted that in the absence of confounding factors, such as interactions with other organisms and land use change, climate warming of order of 1 °C could advance first and peak appearance of most butterflies by 2–10 days (Roy and Sparks, 2000).

***Biological control and temperature*:** Most of the insect predator and parasitoid groups are expected to produce an additional 2–4 generations each year. However, there is still uncertainty regarding the extent to which host-parasitoid phenology synchronized after an increase in generations (Kiritani, 2006). However, there may be some negative effects on biocontrol due to climate change like biological control of grape mealy bug in California. In California, with increases of 2 and 3 °C in average daily temperature, geographic distribution of mealy bug remains relatively unchanged, but areas favorable for its development shift northward with population levels generally increasing everywhere due to increasing failure of biological control. The coccinellid predator *Cryptolaemus montrouzieri* is predicted to be most abundant in south where grape culture is sparse and along the near coastal areas of Central California. The distribution of parasitoid *Anagyrus pseudococi* is quite similar to that of mealy bug, while parasitoid *Leptomastidae abnormis* is most abundant in hotter desert regions of southern California (Gutierrez *et al.*, 2008).

***Aphids*:** Aphid biology is influenced by temperature, which initiates development, fecundity, movement and morphological determination. In *Myzus persicae,* increase of mean temperature by 1 °C advances the time of progeny migration by two weeks, also well evident in *Acyrthosiphon*, *Rhopalosiphum* and *Sitobion*. The impact of temperature on aphid damage and ability of many species to act as vectors are well known. Temperature also contributes to morph determination besides differences in responses between species, between clones within species and within morphs within a clone. Aphids are very sensitive to warmer conditions with a reduced tendency for alate production, lower migratory potential and delay in sexual morph production. So the direct effect of warming on aphids is likely to beneficial in relation to individual performance, with important consequences on population dynamics and their pest status. Various aphid species like *Aphis gossypii*, *Myzus persicae etc*. which are worldwide in distribution, may increase by more number of generation a year due to lower developmental zero point and low thermal totals required for one generation, resulting in more infestation.

***Stink bug*:** The winter mortality of adults of *Nezara viridula* is predicted to be reduced by 15% by each rise of 1 °C in Japan (Kiritani, 2006). Some of the insects in tropics are already living at the limit of their temperature range and any further increases could quickly kill them off with huge repercussions for tropical habitats, which rely on insects for everything from pollination to waste disposal. Recent findings showed that a rise in average temperatures in tropics of just 1 °C or 2 °C could be enough to exert a significant and harmful effect on the survival of a wide variety of important insects. Insects are critical to health of tropical habitats because they perform vital services such as breaking down organic matter, pollinating flowers to produce fruits and nuts and providing sustenance for creatures higher up the food chain. However, it may not be true with many of the pest species. Insects in colder regions may survive and proliferate because they are still at the lower end of temperature. For example, pea aphid sampled at 52 °N, has a critical thermal limit of -10 °C–35 °C, but maximum temperature it has lived in has been 20 °C. In case of a pod-sucking tropical bug sampled at 6 °N, the limit is 15–32 °C and the maximum temperature has already gone up to 32 °C. "In tropics, many species live at or near thermal optimum, a temperature that lets them thrive, once the temperature gets above the optimum, fitness levels will decline quickly".

***Disease vectors*:** Insect vectors transmit many diseases in horticultural crops. Existing studies of climate change on organisms suggest that direct effects of temperature are likely to be larger and more important than any other factor. The main effect of temperature in temperate regions is to influence winter survival of vectors. Natural spread of vectors, pests and diseases is accelerated towards the north, as former climate barriers are no long effective. This results in more severe outbreaks of plant-disease vectors like aphids, whiteflies, thrips, etc., an extension of period of disease infection further into the growing season and also introduction and establishment of new vector species. More vectors survive from one vegetation period to the next leading to earlier and faster development of transmitted diseases. Citrus greening disease is transmitted by psyllid, *Diaphorina citri*. Presently, disease is more prevalent in tropical and subtropical countries in the world. Climate change may push the disease in temperate zones.

***Nematodes*:** There is a close correlation between soil temperatures and distributions of some plant-parasitic species, such as *Longidorus caespiticola*. A 1 °C rise in temperature would allow this species to become established further north in Great Britain. Root-knot nematode, (*Meloidogyne incognita*) was recently found in the Netherlands, a species that previously was considered restricted to the Mediterranean area.

The development of temperate nematode pests such as potato cyst nematode (*Globodera pallida*) and root-knot nematode (*M. hapla*) has been observed on potato roots from localities where these nematodes were absent earlier. An increase of 2-3 °C higher soil temperatures in traditional mild climate areas was observed during the past 20 years, which might have acclimatized the pest to develop under warmer climate (Krishna Prasad, 2003). Thus, increase in soil temperatures at traditional localities may encourage the nematode spread in the newer areas.

Interaction of Elevated CO_2 and Temperature

Doubling the current atmospheric CO_2 concentration, and an increase in global mean annual temperatures of 1.5–6 °C, have been predicted to occur by the end of this century. Whilst the separate effects of CO_2 and temperature on plant-insect interactions have been examined in a number of studies, a few have investigated their combined impact. Johns and Hughes, (2002) have conducted a factorial experiment to explore the effect of a doubling of CO_2 concentration and a 3 °C temperature increase on the development of a complete generation of leaf-miner,

Dialectica scalariella, in the host plant Paterson's Curse, *Echium plantagineum*. Elevated CO_2 increased biomass, reduced leaf N and increased C : N ratios in the host plants. Leaf thickness also increased under elevated CO_2, but only in high-temperature treatment. Female *D. scalariella* did not discriminate between plants grown at the different CO_2 levels when ovipositing, despite the reduction in foliage quality under elevated CO_2. Overall, the negative response of *D. scalariella* to elevated CO_2 was greater than for many species of free-living insects, presumably because of limited mobility imposed by the leaf-mining habit. Development was accelerated at the high temperature and slowed under elevated CO_2.

The net result was a reduction in development time of 14 days in elevated CO_2/high temperature treatment, compared to the ambient CO_2/low temperature treatment. Larval survivorship and adult moth weight were both affected by a significant interaction between CO_2 and temperature. At the low temperature, CO_2 had little effect on survivorship, but at the high temperature, survivorship was significantly reduced under elevated CO_2. Similarly, elevated CO_2 had a stronger negative effect on adult moth weight when combined with high-temperature treatment. Thus, high temperature accelerated insect development to such an extent that larvae did not have sufficient feeding time to compensate for poorer quality of the foliage. The frequency with which interactions between CO_2 and temperature affected both plant and insect performance in this study highlights the need for caution when predicting the effects of future climate change on plant-insect interactions from single-factor experiments.

Effect of Increased Pollutants (ozone and nitrous oxide)

Ozone is likely to have an adverse effect on plant growth. It appears plants are less sensitive to nitrous oxide. Atmospheric pollutants directly or indirectly affect the process of water and nutrient uptake from soil and making the plant weak thereby affecting the insects.

Effect on Host Plant Distribution

Climate change will affect the geographical distribution of plant species and their growth patterns. Spatial changes of crops and cropping systems for sustained productivity an imminent under climate change scenario affecting the distribution of insect pests.

Pest Shifts in Geographical Distributions

Climate change will affect the geographical distribution of plant species, their growth patterns thus affecting the distribution of insect pests. Because of change in temperature regimes pests may spread into newer areas where hitherto are not suitable for insect development and multiplication. This situation can be explained with the help of oriental fruit fly, *Bactorcera dorsalis*, a major pest throughout South East Asia and in a number of pacific Islands. Stephens *et al.* (2007) studied with the help of CLIMEX model the potential global distribution of *B. dorsalis* under current and future climate scenarios. Under the current climatic conditions, its projected potential distribution includes much of the tropics and subtropics and extends into warm temperate areas such as southern Mediterranean Europe. The model projects optimal climatic conditions for *B. dorsalis* in southeastern USA, where the principle range-liming factor is likely to be cold stress. As a result of climate change, the potential global range for *B. dorsalis* is projected to extend further pole wards as cold stress boundaries recede. However, the potential range contracts in areas where precipitation is projected to decrease substantially. Similarly, the pest has been free in many areas in India. With the changing climate scenarios this pest may spread to newer areas on horticultural crops. The significant increases in the potential distribution of *B. dorsalis* projected under the climate change scenarios suggest that the World Trade Organization should allow biosecurity authorities to consider the effects of climate change when undertaking pest risk assessments.

Climatic warming will allow the majority of temperate insect species to extend their ranges to higher latitudes and altitudes. Minimum rather than maximum temperature is important in determining global distribution of insect species.

A study of 35 non-migratory butterflies species in Europe found in recent decades, that about two-thirds have expanded their range by 20–150 miles and many plants flower a week earlier than they did 50 years ago. So as temperatures rise, seasonal cues are altered and life shifts in space and time.

Warmer winter temperatures are likely to increase the probability of the spread of American leaf miner, *Liriomyza* that has been introduced into our county some time back and more prevalent in southern states than northern states.

Climate Change, Migration and Over Wintering of Pests

Migration in insects occurs for escaping adverse conditions. However, with the increase in temperature with climate change, unfavorable areas may become suitable thereby eliminating the need for migration. Any increase in temperature will result in greater ability to overwinter at higher latitudes, ultimately causing poleward extension of insect distribution. For example, cabbage butterfly migrates from hills to plains in winter and back to hills in summer. Global warming may affect this migration and extend distribution of the pest further north. However, the activity of this pest may decline in the plains (Chander *et al.*, 2008). Aphids are major pests on many horticultural crops and are most likely affected by temperature change as a result of their low developmental threshold, low thermal constant and multivoltinism (Kiritani, 1997).

In India, fruit fly, *Bactrocera zonata* is generally restricted to north India. Till late 1990's this fruit fly was over wintering in north India. But during the last 3-4 years adults are being observed during winters in Uttar Pradesh most probably because of increase in soil temperatures due to climate changes (personal discussion with Dr A. Verghese).

With the rise in temperature, onset of hibernation may be delayed while it may be suspended earlier than usual in spring, thereby increasing the active period of pests, which would result in greater damage. Where an insect chooses to over winter may affect its survival. For example, some insects over winter near the soil surface. In this case, snow cover can provide an excellent insulation, and if this is reduced, these insects might not survive a very cold winter. On the other hand, an increase in temperature and a longer growing season could result in an extension of ranges of plants and animals, with new insects being able to establish themselves in a region where they were previously unable to survive.

Migratory pests, in particular locusts, are totally dependent on rain, temperature and vegetation and their habitats change rapidly. The desert locust (*Schistocerca gregaria*), like other locusts, can change its behavior and physiology from solitary grasshoppers to gregarious stages that from swarm. Solitary desert locusts occur at low density in the recession area, which covers North Africa, the Sahel, the Red Sea countries and parts of India, Pakistan, Iran and Afghanistan. The outbreak area reaches from Mauritania to India and from southern Europe to Cameroon and Tanzania. Outbreaks and plagues originate in the recession areas when there are several cycles of good breeding conditions. Although the effects of climate change on this system are difficult to judge, climate scenarios with more winter rain in the Sahel may provide better breeding conditions.

Changes in Interaction Between Species

Rapeseed-mustard is infested by two aphid species, *Liphaphis eysimi* and *Myzus persicae*, the former being dominant during severe winters and the latter during mild winters (Chander and Phadke, 1994). With rise in temperature, higher incidence of *Myzus persicae* may be witnessed. Such faunal shifts may also take place in other crops.

Pest-Natural Enemy Interactions (Including Tritrophic Interactions)

Temperature and rainfall can influence the efficacy and performance of bioconrol agents like parasites, predators and microbials. With the increase in temperature, predators of aphids on *rabi* crops may become more active than at present. Besides, many parasitoids may lose the "tracking" ability for the host caterpillars to deposit the eggs due to unsolicited appearance of heavy rains. Wasps for instance, may lose the critical "window period" or hatching out of cocoons due to late onset of rains. On the contrary, some species of pests like aphids are adversely affected due to fungal outbreaks during the rains. The consequence on population dynamics may lead to substantial increase in the use of agrochemicals and related load on the environment.

Effect of Drought on Insect Incidence

Under the changing climatic conditions drought will have the following effects on the pest status: (i) drought may enhance the incidence of some pests, *e.g.,* termites; (ii) affect the diapause of insect stages. Under drought conditions eggs of grasshoppers may undergo diapause and results in abnormal build-up of pest populations in the next year; (iii) drought stress may result in higher damage to plants due to insects, *e.g.,* leaf miners; (iv) drought and host plant vs insect's species: frequent droughts may lead to complex rapid (increase and decrease in numbers) in ecosystems where different species (butterflies, moths, carabid beetles) of insects are involved and needs to be considered in planning conservation strategies and (v) drought - natural enemies of the pests : drought affects the efficacy of biocontrol agents.

Climate Change and Pesticides

The consequence on population's dynamic may lead to substantial increase in the use of agro-chemicals and related load on the environment. Higher temperature leads to faster completion of generation cycle of pest which indirectly provokes indiscriminate use of pesticide and hence pesticide resistance.

Thus: climate change could affect efficacy of crop protection chemicals through: (i) changes in temperature and rainfall and (ii) morphological and physiological changes in crop plants. Globally, climate change models predict an increase in probability of intense rainfall, which could result in increased pesticide wash off and reduced pest control. In contract, increased metabolic rate at higher temperature could result in faster uptake by plants and higher toxicity to pests. The rates of pesticide application thus have to be modified carefully according to new situations as they arise.

Climate change may also affect ability to control pests. For example, high temperature is reported to reduce the effectiveness of some pesticides. Humidity levels can also modify their efficacy, as can the timing and amount of rain following their application. On a simpler level, rain can affect growers' ability to apply the pesticide at the time of most need, possibly an increasingly likely scenario. If pests are able to complete more generations in a season then this may lead to greater pesticide use, which in turn may lead to the more rapid development of pesticide resistance. Strategies to avoid the development of resistance need to be planned in advance and rely on the availability of a range of pesticides and/or alternative control methods. This scenario is mot likely for some the horticultural crop pests like mites, leaf miners, thrips *etc*.

Mealy bugs are emerging as a major problem in many of horticultural crops. Fish oil rosin soap (FORS) is highly effective against mealy bugs but is found to be phytotoxic on some plants. So with climate change, there is a high risk of FORS phytotoxicity with increasing temperature. The vulnerability of horticultural industries to fruit fly *Bactrocera* sp. under climate change may lead to increased damage and control costs to commercial growers. In Australia, costs to mainland apple, orange, and pear growers are estimated to increase by 25, 38 and 95% with increases of 0.5 °C, 1.0 °C, and 2 °C temperature, respectively (Sutherst, 2000).

Aberrant Weather and Extreme Events

- *Helicoverpa armigera* is appearing in severe proportions once in 5-6 years and resulting in reduced efficacy of insecticides leading failure of IPM and crop.
- Similarly, whitefly, *Bemisia tabaci* menace has been increasing in the last 20 years because of climate change coupled with change in the crop varieties and new agronomic practices. Development of new biotypes has been noticed in whitefly. Incidence of thrips and mealy bugs has been increasing on onion, rose, grape *etc*. in the last 20 years.
- Red spidermite, *Tetranychus urticae*, which was considered as minor pests two decades back, has been regularly appearing on crops like tomato, grape and other ornamental crops.
- Incidence of diamondback moth, *Plutella xylostella* has come down and incidence of *Spodoptera exiqua* and *S. litura* infestation has increased on a number of horticultural crops in the past two decades.
- Thrips transmitted diseases/tospovirus on crops such as watermelon, tomato, chilli and other horticultural crops is a new phenomenon in the last 20 years.
- Climate change during the past eight years has played a critical role in apple pollination failure particularly in Himachal Pradesh. Rains during the flowering season affected the pollination by wind and insects. Low temperatures also adversely affect fruit set in apple.

Adaptation Measures for Climate Change and Global Warming

- Increased surveillance of pest and diseases for taking timely plant protection measure.
- Alterations in sowing dates of crops like vegetables, annual ornamental crops among various horticultural crops to alter host pest synchrony.
- Explore changes in host-pest interaction under early, normal and late sown conditions in order to recommend optimum sowing dates for reduced pest pressure and increased yield of vegetable crops.
- Breeding of cultivars for both pest and drought resistance in different vegetable crops.

- Evolve strains of natural enemies for higher temperature tolerance.

FUTURE STRATEGIES FOR RESEARCH

- Information to be generated on effect of cumulative increase of carbon dioxide and temperatures on biology of insect and mite pests.
- Intensification of research efforts on marker aided selection and transgenic development for biotic stress management.
- Validation of prediction models of horticultural crop pests under changing climate scenarios.
- Studies on change in the pattern of insect transmitted diseases.
- Insecticide resistance management studies on sucking pests, which are likely to increase in the present scenario.
- Development of adaptation and mitigation measures for minimizing the effect of climate change for reducing pest risks.

CONCLUSION

Agriculture and horticulture crops are likely to face growing degree of challenge from pests due to global warming and climate change. The pest problem will be aggravated in several ways to demand matching solutions. Global warming will change the composition of pest fauna in different areas. The length of growing season for tropical and subtropical pests will increase, resulting in more generations and hence greater pest impact. Over wintering of pests will increase producing large populations as a base for build-up in the following seasons. In some cases, pests may not require hibernation thereby producing many more generations than at present. Many pest species may have their diapauses strategies disrupted as the linkages between temperature and moisture regimens and day lengths are alerted. The reproductive ability of pests might be affected by the changes in nutritional value of plants they feed upon due to elevated CO_2 effect under climate change. Some pests will inevitably become a more serious problem, whilst others may decline.

There will be increased variability in pest impact caused due to greater variability in climate. Pest-plant interactions will change and may cause increased damage. Relationship between pest and natural enemies will change affecting the status of pest species. Economics of pest control measures may change affecting benefit : cost (B:C) ratio both due to increased cost of control measures and reduction in crop production. Climate change is likely to alter the balance between insect pests and their natural enemies and their hosts. Biological control using native natural enemies is expected to become a more important control tactic.

Advance contingent plans would be required for aberrant weather conditions along with changes in some policy issues like improved credit availability and extension of crop insurance schemes in rainfed areas.

Plant protection will take new dimensions for tackling these new scenarios. Realistic approach to tackle the growing pest problem would be management strategies involving exclusion or early detection and elimination wherever possible. Quarantine system need to be tightened, because with liberalization of world trade there is increased scope for introduction of pests from other countries and establishing in the imported country under climate change.

References

Ananthakrishnan, T.N. 2007. Insects and Climate. Entomology Academy of India Base paper No. 1. pp.27.

Atwal, A.S. 1955. Influence of temperature, photoperiod and food on the speed of development, longevity, fecundity and other qualities of diamondback moth, *Plutella maculipennis* Curtis (Lepidoptera:Tineidae). *Aust. J. Zool.* **3**:185-221.

Bezemer, T.M., Knight, K.J., Newington, J.E., Jones, T.H. 1999. How general are aphid responses to elevated atmospheric CO_2? *Annals of the Entomological Society of America*. **92**:724-30.

Chander, S. and Phadke, K.G. 1994. Incidence of mustard aphid, *Liphaphis erysimi* and potato aphid, *Myzus persicae* on rapeseed crop. *Ann. Agric. Res.* **15**:385-87.

Chander, S., Aggarwal, P.K., and Prem Kishore, 2008. Food Crisis and Impact of Climate Change on Crop Pests. *Crop Care* **33**:31-39.

Coaklely, S.M., Scherm, H. and Chakraborty, S. 1999. Climate change and plant disease management. *Ann. Rev. Phytopathhol.* **37**:399-26.

Guerenstein, P.G. and Hildebrand JG. 2008. Roles and Effects of Environmental Carbon Dioxide in Insect Life. *Annu. Rev. Entomol.* **58**:161-78.

Gutierrez, A.P., Ponti, L., Oultremont d ,T., and Ellis. C.K. 2008. Climate change effects on poikilotherm tritrophic interactions. *Climate Change* **87 (Suppl 1):** S167-S192.

John, C.V. and Hughes, L. 2002. Interactive effects of elevated CO_2 and temperature on the leaf-miner *Dialectica scalariella* Zeller (Lepidoptera: Gracillariidae) in Paterson's Curse, *Echium plantagineum* (Boraginaceae). *Global Change Biology* **8**:142-52.

Kirtani, K. 1997. The low development threshold temperature an thermal constant in insects, mites and nematodes in Japan. Miscellaneous Publications of the National Institute of Agro-Environment Sciences **21:**1-72.

Kirtani K. 2006. Predicting impacts of global warming on population dynamics and distribution of arthropods in Japan. *Population Ecology* **48:**5-12.

Krishna Prasad, K.S. 2003. Intensity and buildup of potato cyst nematodes in varying ecological situations at Nilgiri hills. *J. Indian Potato Assoc.* **30**:157-58.

Lincoln, D.E., Couvet D and Sionit N. 1986. Response of an insect herbivore to host plants grown in carbon dioxide enriched atmospheres. *Oecologia* **69:**556-60.

Osbrink, W.L.A., Trumble JT and Wagner RE. 1987. Host suitability of *Phaseolus lunata* for *Trichoplusia ni* (Lepidoptera: Noctuidae) in controlled carbon dioxide atmospheres. *Environmental Entomology* **16:**639-44.

Patterson, D.T., Westbrook, J.K., Joyce R.J.V., Lingren, P.D., Rogasik, J. and Reilly, J.M. 1999 Weeds, insects, and diseases. Special issue. Climate change: impacts on agriculture. *Climatic-Change* **43:**711-27.

Roy, D.B. and Sparks, T.H. 2000. Phenology of British butterflies and climate change. *Global Change Biology* **6:**407-16.

Stephens, A.E.A., Kriticos, D.J. and Leriche, A. 2007. The current and future potential geographical distribution of the oriental fruit fly, *Bactrocera dorsalis* (Diptera : Tephritidae). *Bulletin of Entomological Research* **97:**369-78.

Sutherst, R.W., Collyer, B.S. and Yonow, T. 2000. The vulnerability of Australian horticulture to the Queensland fruit fly, *Bactrocera (Dacus) tryoni*, under climate change. *Australian Journal of Agricultural Research* **51:**467-80.

Yamada, H. and Kawasaki, K. 1983. The effect of temperature and humidity on the development, fecundity and multiplication of the diamondback moth, *Plutella xylostella* (L.). *Japanese J. Appl. Entomol. Zool.* **27:**17-21.

Yamaguchi T, Kiritani K, Matsuhira K and Fukuda K. 2001. The influence of unusual hot weather on the occurrence of several arthropod crop pests.Japanese *J. Applied Entomology and Zoology* **45:**1-7.

22

Impact Assessment of Climate Change on Potato Diseases and Insect Pests

B. P. Singh and Narayana M. Bhat

METEOROLOGICAL conditions are known to affect crop pests and pathogens. The range of plant pathogens and insect pests are mainly constrained by temperature, and the frequency and severity of weather events affecting the timing, intensity and nature of outbreaks of most organisms (Yang and Scherm, 1997). Milder winters and warmer nights allow increased winter survival of many plant pests and pathogens, accelerate vector and pathogen life-cycles, and increase sporulation and infectiousness of foliar fungi. Climate change and global warming will allow survival of plants and pathogens outside their existing geographical ranges. Northward and southward range shifts in insect pests and diseases with warming is indicated in the Northern and Southern Hemispheres, respectively (Sutherst, 1990; Coakley *et al.*, 1999). Frequency of extreme events of droughts and flooding is also likely to increase with global warming, thereby creating conditions favoring fungal infestations of leaf, root and tuber crops in flooding and runoff areas.

The three legs of the "plant disease triangle" namely the host, pathogen, and environment must be present and interact appropriately for plant disease to appear. If any of the three factors is altered, changes in the progression of a disease epidemic can occur. The climate change and global warming with increases in temperature, moisture and CO_2 levels can impact all three legs of the plant disease triangle in various ways.

IMPACT ANALYSIS ON POTATO DISEASE AND INSECT PESTS

Effect of Elevated CO_2

Plant diseases: Not much information has been generated on the impact of elevated CO_2 on potato diseases. However, a general outline has been drawn from the available literature on the effect of elevated CO_2 on plant diseases. The concentrations of CO_2 in troposphere are projected to increase from 355 ppm (*v/v*) to 710 ppm by the year 2050 (Manning and Tiedemann, 1995).

It is expected that higher CO_2 concentrations will result in increased plant biomass production due to increased photosynthesis, where there is sufficient soil moisture. The increase in CO_2 will probably have little direct effect on most pathogens, as many soil inhabiting fungi can tolerate more than 10- or 20-fold increases in CO_2, and might even be slightly stimulatory (Manning and Tiedemann, 1995). Chakraborty *et al.* (1998) suggest that the impact of increased CO_2 concentrations on plant diseases will likely be through changes in host physiology and anatomy, such as lowered nutrient concentration, greater accumulation of carbohydrates in leaves, more waxes, extra layers of epidermal cells and increased fiber content, and greater number of mesophyll cells. They further reported that two important trends have emerged in the effects of elevated CO_2 on host–pathogen interactions: (1) initial establishment of a pathogen may be delayed because of modifications in pathogen aggressiveness and (or) host susceptibility and (2) increased fecundity of pathogens. The combination of increased fecundity and a more humid microclimate within dense crop canopies associated with increased CO_2 concentrations might provide more opportunities for severe infection. Pathogen growth can be affected by higher CO_2 concentrations resulting in greater fungal spore production. However, increased CO_2 can result in physiological changes to the host plant that can increase host resistance to pathogens as well. Aphid abundance was enhanced by higher levels of CO_2 but parasitism rates remained unchanged (Bezemer *et al.*, 1998).

Insect pests: Work on the effect of elevated CO_2 on potato pests is scanty. Studies carried out by Bezemer *et al.* (1998) revealed that aphid abundance was enhanced by both CO_2 and temperature. Parasitism rates remained unchanged in elevated CO_2 but showed an increasing trend in conditions of elevated temperature.

Effect of Increased Temperature

Late blight: Earlier onset of warm temperatures could result in an early appearance of late blight disease in temperate regions with the potential for more severe epidemics and increased number of fungicide applications needed for its control. Studies carried out in Finland predicted that for each 1 °C warming, late blight would occur four to 7 days earlier, and the susceptibility period extended by 10–20 days (Kaukoranta, 1996). This would result in 1–4 additional fungicide applications, increasing both cost of cultivation and environmental risk. In India also, late blight scenario would change drastically with climate change. Currently, late blight is not a serious problem in autumn in the state of Punjab, Haryana and parts of Uttar Pradesh, primarily due to suboptimal temperature regimes during December-January.

However, disease outbreaks will become more intense with increase in ambient temperature coupled with high RH. Such scenarios have been witnessed during 1997-98 and 2006-07, when average crop losses in this region exceeded 40% resulting in acute shortage of potatoes in the country. Increase in ambient RH coupled with optimum temperature would have a far greater impact of late blight outbreaks in the country. Madhya Pradesh, Gujarat and Central Uttar Pradesh which are comparatively free from late blight attack may witness frequent outbreaks of the disease.

Increase in both, temperature and RH has added new dimension to late blight across the world. Under such a situation *P. infestans* attacks potato stems more often than foliage. In fact, in recent years it is more of 'stem blight' than the foliar blight. This phase of disease is more serious than the foliar stage as it affects the very crop plant. With the change in climate, late

Blight forecasting models will also need to be redefined for different regions which will entail a lot of research work. Bourgeois *et al.* (2004) studied the impact of climate change on disease incidence using mathematical modeling. Disease forecasting systems for late blight, apple scab and Cercospora blight of carrot were evaluated for their relevance in predicting future events related to climate change. It was observed that many of the models developed few years ago are in need of updating. Forecasting systems based on non-linear responses to temperature and leaf wetness will be more accurate in representing the impact of climate change on disease epidemiology.

Early blight: High temperature and RH favour its outbreak. Such situation mostly exists in plateau and sub-mountainous regions, although, it is known to infect potato crop throughout Indo Gangetic plains as well as in high hills. The disease is most favored by temperature ranging from 25–30 °C with an optimum of 26 °C (Dutt, 1979). Prevailing temperatures in the foothills and plateau are already high, further warming may not lead to any substantial effect on the disease.

Soil-borne pathogens: Some of the important soil and tuber borne pathogens are *Synchytrium endobioticum* (Wart), *Spongospora subterranea* (powdery scab), *Rhizoctonia solani* (black scurf and canker), *Streptomyces scabies* (common scab), *Sclerotium rolfsii* (sclerotium wilt), *Macrophomina phaseolina* (charcoal rot), *Rolastonia solanaelarum* (bacterial wilt) and *Fusarium* spp., (wilt and dry rots).

Some of them, *viz., S. endobioticum* causing wart and *S. subterranea* responsible for powdery scab are favoured by low temperature and high soil moisture, whereas rest are favoured by moderate to high temperature and low soil moisture. Consequently, the effect of climate change would also vary from pathogen to pathogen. Wart spores although can cause infection in the range of 10 to 28 °C with an optimum of 21 °C but, there is hardly any infection beyond 23 °C. Therefore, warmer climates are likely to reduce wart infestation. Powdery scab infestation is also likely to be reduced with increase in temperature and reduction in rainfall as a consequence of global warming. Since optimum temperature for powdery scab is 12 °C, and moisture requirement is 100%, the global warming may either lead to elimination of this disease or it will be pushed to higher altitudes making high hills (2500 masl) free of powdery scab.

Diseases like *Sclerotium* wilt, charcoal rot and bacterial wilt are favoured by high temperature and moisture. *Sclerotium* wilt in India is restricted to plateau regions (Madhya Pradesh, Karnataka, Maharashtra). Optimum temperature requirement for this disease is 30–35 °C. With the increase in temperature due to global warming, the disease may enter into other areas like mid hills, and in long run, it may also become prevalent in eastern Indo-Gangetic plains. Similarly, bacterial wilt may also advance to higher altitudes in hilly regions due to global warming making them unfit for seed production.

Charcoal rot is currently endemic in eastern Uttar Pradesh, Bihar and Madhya Pradesh The global warming is likely to increase the severity of this disease in these regions. It is also likely to expand to other parts of North Central plains as well.

Black scurf and common scab are favoured by moderate temperatures (15–21 °C and 20–22 °C, respectively) and are likely to remain insulated from global warming in near future. By the end of the century when ambient temperatures are likely to increase by 1.4–5.8 °C, the severity of these two diseases may decrease substantially.

Viral diseases: Viruses are bane to potato production world over, but more so in sub-tropics and tropics. The rate of multiplication of most of the potato viruses gets increased with increase in temperatures. By the end of this century, average British temperatures are expected to increase by about 3 °C which would help in flaring up of viral diseases.

In India, similar situation may occur in temperate highlands. At present, these hilly regions are considered best for seed potato production but it may not be the case after 2-3 decades. Suptropical plains, where majority of potatoes are grown, represents a different scenario. Although, global warming may not affect potato viruses directly, but may have a serious repercussion through the altered biology of insect vectors. The increase in temperature will enhance vector population thereby increasing the number of insecticide sprays for keeping the vector population in check. Besides rate of multiplication of the virus in host tissue will increase substantially leading to early expression of the virus symptoms and heavy reduction in crop yields.

Insect pests: Findings from Peru suggest that global warming is already affecting the distribution of crop insect pests. Scientists from CIP and the partner institutions are of the opinion that current balance of insect populations is almost certain to be upset. In the opinion of Dr. Aziz Lagnoaui, entomologist from CIP, Lima, Peru, climate change will favour invaders over native species. Some insect pests will become more important while others will decline, but the net effect will be increased pressure of pests on the crop yields. Experiences from Canete Valley in Peru revealed that when white flies are few in number, their feeding on crops does little damage. But if sprayed, they soon develop resistance and multiply. At high population densities, white flies can devastate whole crop stands.

Until recently, only biotype A of *Bemisia tabaci* which causes only moderate yield loss was found in Canete Valley. In 2001, however, a more aggressive whitefly biotype B, was identified. According to farmers, it arrived during El Nino year when climate in Peru was hotter and drier than usual. But instead of disappearing when weather reverted to normal, the pest apparently adapted to its new environment, where it is now inflicting severe damage to crops. Besides, a new species of whitefly, *viz., B. afer* was also detected in the valley. This species is even more aggressive than biotype B of *B. tabaci.* Now it is causing severe crop losses in the valley.

Research findings published in New Scientist by British researchers revealed that farmers in temperate world could be plagued by a growing army of insecticide resistant aphids. All that is holding them back so far has been 'frosty winters', but this natural check on their number could be removed as average winter temperatures rise. Research further revealed that clones of *Myzus persicae* have emerged that are resistant to three major groups of insecticides. Fortunately, these 'super-aphids' are still in minority since their survival is less in frosty nights as compared to sensitive aphids. It is feared that global warming could remove the brake that winter now places on the numbers of insecticide-resistant aphids.

Aberrant Weather and Extreme Events

- ***Increased attack of whitefly***: In India, *B. tabaci* was a minor pest till recently. Data on population build-up during the last 20 years revealed that average population of *B. tabaci* was 11 whitefly/100 leaves during 1984 which rose to 24.24 in 2004. During this period, average ambient temperature increased by 1.07 °C. This indicates that warming may lead to whitefly infestation in Indo Gangetic plains. Increase in *B. tabaci* population led to outbreak of a new viral disease known as apical leaf curl in potato which has since been identified to be caused by a gemini virus. Gemini viruses are not reported to infect potato crop world over. Therefore, a new dimension has been added to seed potato production in sub-tropics. The infestation of this virus in some of the varieties, *viz.*, Kufri Pukhraj, Kufri Anand, Kufri Sutlej reached almost 100% in seed stocks. Consequently, an altogether new approach was adopted to manage this virus which included discarding of the old stocks and their replacement with disease-free stocks from tissue culture and integrated management of whitefly which included delayed planting, seed treatment and weekly application of insecticides including summer oil. This disease has since been contained but the danger persists.

- ***Increased attack of leaf hopper****: Empoasca fabae* (leafhopper) is another pest which has assumed significance in early planted crop in subtropical plains of India. Its population during 1984 was 16.6 which rose to 23.8 in 2004. The hopper burn damage also increased from 45–68% during this intervening period. Sudden warming up of temperature may also lead to flaring up of the pest. This has happened during 2006-07 in Gujarat state. Temperature profile of the region revealed that during crop period right from December to first week of January, ambient temperature was higher and relative humidity was lower than previous years. The increase in temperature was up to 5.8 °C during fourth week of December 2006. This increase in temperature led to faster multiplication of the pest and consequently very heavy (up to 100%) hopper burn damage (Singh *et al.*, 2007).
- ***Increased attack of mites***: Mite infestation has also increased in early planted crop. During 1984-85 its damage was 86% which increased to 100% in 2004.
- ***Increased attack of aphids***: Results tend to suggest that in subtropical plains of India, *Myzus persicae* population is on the rise. During 1984-85, mean aphid population/ 100 compound leaves was 567 which increased to 653 in 2003-04. Besides aphid appearance had advanced by 5 days during last 20 years, reducing the seed production window from 80 to 75 days. Further reduction in the seed production period may adversely affect the seed production programme. On the other hand, population of *Aphis gossypii* has increased three fold during the last 20 years. Although *A. gossypii* has low vector efficiency, its appearance right from the emergence of the crop and further maintaining its population throughout the crop season may pose serious problems to seed production in subtropical plains.
- Studies carried out in Holland revealed that during 12 years (1994–2008), some new viral strains (PVYntn, PVYnw) have been detected indicating that climate change may introduce new viral strains which will have an adverse effect on potato crop.

Adaptation Measures for Climate Change and Global Warming

- ***Alteration in planting date and integrated pest management*** (IPM): Although changes in climate are almost certainly happening, it is not precisely understood how these changes will affect crops, insects, diseases, and their inter-relationships. It is likely that new pests will become established in regions hitherto unknown earlier. The classic example is introduction of new viral disease, apical leaf curl virus hitherto unknown in subtropical plains of India. Adaptation in the form of adjustment in planting date, i.e., delayed planting by 10–15 days along with Integrated Pest Management (IPM) practices has led to the management of apical leaf curl virus quite effectively in subtropical plains. Leafhopper and mite have assumed significance in early planted crop in subtropical plains of India. Its population is on the rise during the last 20 years. Similarly, aphid population has increased in the autumn crop during the last 20 years. Depending on the infestation severity and locality, adjustment of planting date remains an important component of management of insect pests in potato crop. However, this depends on the preceding crop, water availability, diseases and markets. With expectation of introduction of new pests and diseases it is important to have diagnostic tools to detect new pathogens that might be harboured during introductions. Make best use of IPM such as field monitoring, pest forecasting, record keeping, and choosing economically and environmentally sound control measures to be successful in dealing with the effects of climate change.

- ***Change of cultivars and cultural practices***: Selection of suitable cultivars and modification in cultural practices are under climate change. The rate of change in agricultural practices is expected to accelerate with changes in climate. Owing to climate change, drier soils and irregular precipitation would have resulted in increased use of irrigation. Water conservation systems would contribute to reduced leaf wetness and humidity in the crop canopy and, therefore, contribute to reduced foliar diseases compared with more traditional practices such as furrow irrigation. This is particularly true in case of late blight of potato which is highly favored by high humidity and precipitation. Due to climate change, environment stress is expected to rise in the coming years. Therefore, increased emphasis should be placed on breeding cultivars for environmental stress tolerance. New techniques that enhance the identification and development of host tolerance or resistance to biotic and abiotic diseases will be important in facilitating this adaptation.

FUTURE STRATEGIES FOR RESEARCH

- Development of virus resistant varieties.
- Virus-vector relationship.
- Development of early bulking and heat tolerant varieties so that planting can be preponed by 7–10 days to avoid high vector activity and neutralizing the early appearance of vectors.
- Replacement of conventional seed production by tissue culture based seed production.
- Ecology and biology of white fly in relation to apical leaf curl virus.
- Development of late blight resistant varieties.
- Development of disease forecasting system in tune with the climate change.
- Rescheduling of chemical sprays based on new emerging pathogen population.
- Improvement in tools for studying late blight risk and climate change.
- Development of IPM strategies.

CONCLUSION

The information on effect of elevated CO_2 under climate change on pests and diseases is scanty. The elevated CO_2 will probably have little direct effect on most pathogens, as many soil-inhabiting fungi can tolerate more than 10- or 20-fold increases in CO_2, and might even be slightly stimulatory. The impact of increased CO_2 concentrations on plant diseases is likely to be through changes in host physiology and anatomy. Global warming is likely to increase the incidence of viral, late blight, charcoal rot and bacterial wilt, have little effect on early blight and may decrease wart, powdery scab, black scurf and common scab diseases in Indo-Gangetic plains.

There has been perceptible increase in pests during last 20 years to the detriment of crop yield. Increased frequency of aberrant weather and extreme events in the recent past resulted in unexpected outbreak of pest and disease hitherto unknown to a given region. Sudden outbreak of leaf hopper burn in Gujarat during 2006-07, late blight in severe epiphytotic form in Karnataka and Maharashtra during 2006-07 and 2008 and apical leaf curl virus infestation in Indo-Gangetic plains are reminders of climate change and its adverse effect on potato crop. It is therefore, imperative that weather/climate changes are monitored on regular basis along with disease and insect pest infestation. Based on this information, non-linear models need to be developed for predicting effect of climate change on a given disease/pest. Devising adaptation measures including development of resistant varieties for containing the disease and insect pest has become necessary. The insect-pest and vector biology is expected to change dramatically with increase in temperature which needs to be studied and catalogued for the management.

References

Bezemer, T.M., Jones, T.H. and Knight, K.J. 1998. Long term effects of elevated CO_2 and temperature on populations of the peach potato aphid *Myzus persicae* and its parasitoid *Aphidius matricariae*. *Oecologia*. **116**:128-35.

Bourgeois, G., Bourque, A. and Deaudelin, G. 2004. "Modelling the impact of climate change on disease incidence: a bioclimatic challenge." *Canadian Journal of Plant Pathology* **26**:284-90.

Chakraborty, S., Tiedemann, A.V. and Teng, P.S. 1998. Climate change and air pollution: potential impact on plant diseases [online]. *In:* 7th International Congress of Plant Pathology. 9–16 August 1998, Edinburgh. Vol. 1. *Compiled by* British Society for Plant Pathology, Birmingham.

Coakley, S.M., Scherm, H. and Chakraborty, S. 1999. Climate change and plant disease management. *Annu. Rev. Phytopathol.* **37**:399-26.

Dutt, B.L. 1979. *Bacterial and Fungal Diseases of Potato*, ICAR, New Delhi, pp. 141-52.

Kaukoranta, T. 1996. Impact of global warming on potato late blight: risk, yield loss, and control. *Agric. Food Sci. Finl.* **5**:311-27.

Manning, W.J., and Tiedemann, A.V. 1995. Climate change: potential effects of increased atmospheric carbon dioxide (CO2), ozone (O3) and ultraviolet-B (UV-B) radiation on plant diseases. *Environ. Pollut.* **88**: 219-45.

Singh, B.P., Kamlesh Malik, Patel, N.H., Patel, P.J., Singh, S.V. and Sukhwinder Singh. 2007. First report of "hopperburn" damage on potato in Gujarat state. *CPRI Newsletter* **37**:3-4.

Sutherst, R.W. 1990. Impact of climate change on pests and diseases in Australasia. *Search* **21**:230-32.

Yang, X.B. and Scherm, H. 1997. El Niño and infectious disease. *Science* **275**:39.

23

Climate Change and Carbon Sequestration in Horticultural Crops

Debasish Mahapatro

Climate change has been described as a huge threat to mankind in recent times. Since the advent of industrial revolution in mid-1800's, there has been a shift in global atmospheric temperatures. Widespread felling of trees, reduction of agricultural activities, increased use of power dependent activities in agriculture, intemperate use of water resources, etc., have all further accentuated the climate change process.

Mitigating climate change is the need of the hour. Afforestation/reforestation is advocated for mitigation of climate change by locking excess free air CO_2 through carbon sequestration in the biomass of perennial forest trees. However, economic and financial viability of the standalone afforestation/reforestation projects are doubtful, unless these measures supplement or serve as an additional source of income to the people. In contrast, establishment of horticultural orchards, plantations and gardens could be economically highly viable carbon sequestration measures to be adopted on a large scale. Thus, in future climate change scenario horticultural crops particularly the perennial plantation and orchards have high potential for carbon sequestration to capture CO_2 from atmosphere for mitigation of climate change.

CARBON SEQUESTRATION IN HORTICULTURAL CROPS

Carbon sequestration is the process through which agricultural and forestry practices remove carbon from the atmosphere. Sequestration slows the rate of climate change by enhancing carbon storage in trees and soils. Afforestation and reforestation can be an important agent for carbon sequestration. Plants derive their food by the process of photosynthesis. Photosynthesis is a series of enzyme catalyzed reaction which converts light energy into chemical energy. It uses water and CO_2 as substrates, converting them into glucose. The glucose thus produced forms the base of food chain pyramid, with life supporting oxygen as a byproduct. This reaction utilizes six molecules of CO_2 per reaction, hence leading to carbon sequestration. This glucose is utilized for producing edible sugars and plant's supporting structure.

Carbon sequestration is defined by the US Environmental Protection Agency (EPA) as the process through which carbon dioxide (CO_2) from the atmosphere is absorbed by trees, plants and crops through photosynthesis, and stored as carbon in the biomass (tree trunks, branches, foliage and roots) and soils. The term "sinks" is also used to refer to forests, crop lands and grazing lands, and their ability to sequester carbon. Agriculture, horticulture and forestry activities can also release CO_2 to the atmosphere. Therefore, a carbon sink occurs when carbon sequestration is greater than carbon releases over some time period.

There is a large and growing interest in carbon sequestration as a method of combating climate change. Agriculture has an immense adaptive capacity for carbon sequestration. Farm activities can sequester carbon and also reduce emissions of methane and nitrous oxide in a variety of ways. These include (to name a few) conservation tillage, conservation or riparian buffers, grazing land management, reducing the use of synthetic fertilizer, handling manure differently, capturing and burning methane, reducing irrigation, and switching to biofuels.

CARBON FOOTPRINTS OF HORTICULTURE AND SCOPE FOR MITIGATION

The huge potential of horticultural crops for carbon sequestration can be truly estimated only if we understand the carbon footprints of horticulture to reduce emissions as well, through various methods. Modern horticultural practices have led to rampant usage of fertilizers, pesticides and machinery. These themselves are energy intensive leading to huge emissions of greenhouse gases. Hence, we need to understand and minimize activities allied to agriculture which contribute vastly to emissions. We need to understand carbon footprints of each allied activity and make appropriate decisions accordingly. Stress must be given on activities with low carbon emission, replacing high polluting practices.

Fertilizer

Fertilizers are the biggest source of greenhouse gases with reference to agriculture. A large amount of CO_2 and N_2O is released into atmosphere during the production of fertilizers; offsetting the amount of carbon sequestered by trees. The N_2O has more adverse effect than CO_2 over global warming, since each tonne N_2O equivalent to 300 tonnes of CO_2. Fertilizer production is an energy-intensive process, utilizing huge amounts of fossil fuels such as coal, natural gas and naptha.

Fertilizers are generally transported over large distances of up to 1,000 km which further leads to CO_2 emission. The transport of fertilizer adds to the footprint as a tonne of fertilizer transported over 1 km releases about 1 kg of CO_2.

A large amount of $CO_2/N_2O/CH_4$ is also released due to different chemical interactions between soil, moisture and fertilizers. Addition of nitrogen to soil in any form (animal excreta, synthetic fertilizer, crop residues or biological fixation), results in increased nitrous oxide emissions. A study revealed that bulk of nitrogen added to New Zealand soils comes from the excreta of animals (1282 Gg N/y) and addition of fertilizer (213 Gg N/y). On an average, only 10.5% of nitrogen in grass, silage or other feedstuff is converted by grazing animals into milk, meat, eggs or wool, and the remainder is excreted in dung and urine.

Nitrous oxide is emitted directly from soil as a result of these inputs. It is also generated indirectly when nitrate lost by leaching or run-off is converted to nitrous oxide in water bodies, and when nitrogen from excreta and fertilizer is lost as ammonia to the atmosphere and subsequently deposited on land. During 2000, nitrous oxide emission from these sources amounted to 25.2 Gg N, compared with the 1990 base value of 24.4 Gg N.

Decay of Biomass

There is another aspect of horticulture which often goes unnoticed and unmonitored. It is the aspect of biomass decay through anaerobic fermentation of dried leaves and other biomass which leads to formation of methane, which is a potent GHG. It is difficult to estimate the exact amount of methane generated due to vastly erratic nature of processes involved as well as differences in climate and aerobic conditions. While it is difficult to eliminate, one possibility is to promote the use of biomass in a controlled way for heat generation which ensures that biomass is converted to lower potency CO_2 or to use it in vermincomposting farms for conversion to organic manure which has multiple benefits. Usage of biomass as a bio fuel or as a vermin compost would not only avoid methane production but also displace the more energy consuming fertilisers thus reducing the overall carbon footprint of the cultivation process.

Applying organic fertilizers, such as those resulting from composting, to agricultural land could increase the amount of carbon stored in these soils and contribute significantly to the reduction of greenhouse gas emissions. Biomass could be used for the following possibilities:

- Reduction of dependence on fossil fuels for energy through large-scale energy plantation which can be harnessed for energy generation.
- Use of biomass waste for heat generation.
- Use of biomass waste for composting and hence on substitution of synthetic fertilizers.

HORTICULTURE AND CLEAN DEVELOPMENT MECHANISM (CDM)

Horticulture provides an economically viable CDM. Because, perennial horticultural crops like plantation fruit and orchard crops can effectively supplement or generate additional source of income to attain economic and financial viability of CDM project.

The CDM, an arrangement under the Kyoto Protocol allows industrialized countries with a greenhouse gas reduction commitment to invest in projects that reduce emissions in developing countries as an alternative to more expensive emission reductions in their own countries. Conceived by UNFCCC along with IPCC and supported by more than 180 countries, CDM has proven to be a boon for adoption of cleaner and efficient technologies. It has acted as an impulse to otherwise unviable projects.

It encourages carbon mitigating and sequestering methodologies around the globe. Its objective is to achieve "stabilization of greenhouse gas concentrations in atmosphere at a level that would prevent dangerous anthropogenic interference with the climate system". An incentive in the form of Carbon Emission Reduction (CER) or Carbon Credits are issued to project proponents for enabling reduction in greenhouse gas emission.

One CER can be termed as 1 tonne of CO_2 reduced which would have been produced in absence of project activity. Every CDM project follows a fixed process and a methodology for evaluating and estimating CER generation. As per the UNFCCC guidelines, afforestation and reforestation is also an applicable CDM project and it can avail the CDM benefits.

Additional CDM Opportunities in Horticulture

Afforestation and reforestation holds immense CDM potential. It not only acts as a carbon sink, but can also help in replacing energy intensive fertilizers and fossil fuel.

The CDM opportunities are:

- Replacement of fertilizers by organic compost.
- Use of biomass as fuel in cogeneration plants.
- Restoration of degraded land.

Replacement of Fertilizers by Organic Compost

Fertilizers are very energy intensive during manufacturing processes and usages and one of the biggest sources of greenhouse gases. Organic agriculture achieves high plant yields by making efficient use of organic residues: to enrich soils, it uses composted harvest residues and animal manure. This saves 50–150 kg, depending upon the crop, in synthetic nitrogen fertilizer per hectare which would otherwise need to be produced using non-renewable fuels.

Studies have shown that conventional arable farming operations in England consume some 17,000 litres of fossil fuels embodied in fertilizers/100ha of land each year. Worldwide, 90 million tonnes of mineral oil or natural gas are processed to nitrogen fertilizer every year. This generates 250 million tonnes of CO_2 emissions.

With their low-impact methods, organic farmers boost soil fertility and humus content of soils. The result is that the greenhouse gas CO_2 is returned to the biomass of soil. Long-term field trials conducted over many years in Switzerland have shown that compared to other methods of farming (conventional, integrated production) organic farming enriches 12–15 percent more carbon dioxide in the soil. This means that organic farms return 575–700 kg CO_2 to soil/hectare/year more than other farms. Scientists in Austria and Germany have confirmed this figure. Organic farming thus reduces CO_2 emissions by eliminating synthetic fertilizers, and at the same time reduces atmospheric concentrations of this gas by storing it in the soil – a true win-win strategy.

A further valuable outcome of organic farming method is that soils with higher humus content can adapt better to adverse effects of climate change. Soils rich in humus store more water and for longer period, which is particularly effective in ensuring higher yields during lengthier dry periods in the summer. The improved water retention potential of soils also protects against sudden and strong rainfall, as rivers rise less rapidly and runoff erosion is slowed.

Assuming 1 tonne of organic compost would replace 0.1 tonnes of fertilizer, and assuming all other conditions remain equal, 1.5 CERs would be generated due to production of 1 tonne of organic compost which at present prices would fetch about Rs.1,500.

Around 170 kg/ha/annum fertilizer is used in horticultural practices, implying that for a land of 1,000 ha around 1,700 tonnes of fertilizer is used. Even, if we substitute 30% of fertilizer by organic compost, we can generate up to Rs.7,650,000/annum.

Use of Biomass as Fuel in Cogeneration Plants

Industrial boilers, a prime source of generation of steam in India utilizes fossil fuels such as coal. As a result of which, a huge amount of harmful greenhouse gases (GHG) along with other toxic gases are emitted in the environment. It results in increase in generation of greenhouse gases and flyash. Flyash, a combustion product of coal is very harmful in nature. Inhalation of fly ash results in tuberculosis, bronchitis, pneumoconiosis and other harmful respiratory diseases.

Biomass such as wood chips can replace commonly used fuels such as coal or natural gas as fuel in power plants. Every 1 MW of power plant requires around 4,500 tonnes of biomass and generates CERs which could be worth up to Rs 7,000,000 per annum.

Restoration of Degraded Land

Afforestation of degraded land through horticultural practices not only improves land quality but also results in increased biodiversity and carbon sequestration. Afforestation can generate the income to the poor farmers/communities by enabling the carbon sequestered by plantations to act like a "virtual cash crop" for the local project beneficiaries, who will gain direct benefits from harvesting the plantation as well as from the sale of carbon credits. In addition, forest restoration in this area plays a vital role in biodiversity conservation, soil and water conservation and poverty alleviation, while sequestering carbon dioxide from the atmosphere.

A share holding agreement between landowners, investors and other stakeholders can be framed to facilitate horticultural practices on degraded land. Up to 50 CERs/ha/year can be generated by plantation of fast growing species.

Carbon Sequestering Potential of Soils

Carbon sequestration in soil has been recognized by the Intergovernmental Panel on Climate Change and the European Commission as one of the possible measures through which greenhouse gas emissions can be mitigated.

One estimate of the potential value of this approach – which assumed that 20% of the surface of agricultural land in EU could be used as a sink for carbon – suggested it could constitute about 8.6% of the total EU emission-reduction objective. An increase of just 0.15% in organic carbon in arable soils in a country like Italy would effectively imply the sequestration of the same amount of carbon within soil that is currently released into atmosphere in a period of one year through the use of fossil fuels. Furthermore, increasing organic matter in soils may cause other greenhouse gas-saving effects, such as improved workability of soils, better water retention, less production and use of mineral fertilizers and pesticides, and reduced release of nitrous oxide.

Dedicated measures are required to enhance soil sequestering capacity. Composting can contribute in a positive way to the twin objectives of restoring soil quality and sequestering carbon in soils. Applications of organic matter (in the form of organic fertilizers) can lead either to a build-up of soil organic carbon over time, or a reduction in the rate at which organic matter is depleted from soils. In either case, overall quantity of organic matter in soils will be higher than using no organic fertilizer. Organic fertilizers can reverse the decline in soil organic matter that has occurred in relatively recent decades by contributing to the build-up in the stable organic fraction in soils, and having the effect, in any given year, of ensuring that more carbon is held within the soil.

However, calculating the value of this technique to climate change policies is complicated. To refine previous calculations and to take account of the positive and negative dynamics of carbon storage in soil, the dynamics of compost application and build-up balancing this with mineralization and loss through tillage has been modeled. Results suggest that soils where manure was added have soil organic carbon levels 1.34% higher than un-amended soils, and 1.13% higher than soils amended with chemical fertilizers, over a 50-year period.

CONCLUSION

It is easily demonstrated that agriculture and horticulture play very strong roles in determining the emission levels of GHGs in atmosphere. Hence, it devolves onto the scientific community to propagate practices, which encourage conscience backed green cultivation practices that contribute to not only economic growth of horticulturalists, but also ensure environment quality. In future, climate change scenario in horticultural crops, particularly the perennial plantation and orchards have high potential for carbon sequestration to capture CO_2 from atmosphere for mitigation of climate change.

24

Impact Assessment and Adaptation Strategies to Mitigate the Effect of Climate Change on Island Ecosystem

S. K. Ambast and R. C. Srivastava

COASTAL and island ecosystems are intrinsically linked to climate in many ways. The ocean is an important distributor of the planet's heat, with major ocean currents moving heat towards the poles from the equator. There is some chance that this distribution of heat through the ocean's "conveyor belt" circulation would be strongly influenced by the changes projected in many global climate models. Sea-level rise is another climate-related phenomenon with a major influence on coastlines and islands. Global sea-level has already risen by 10–20 cm in the past century and the best estimate suggest that sea-level will rise by an additional 48 cm by 2100, with an uncertainty range of 13–95 cm. Accelerated global sea-level rise is expected to have dramatic impacts in islands, where subsistence and erosion problems already exist.

Developing, least developing and islands states/nations are more vulnerable and will face additional challenges to attain the United Nations Millennium Development Goals by 2015 (UN, 2007). India has been identified as one amongst 27 countries which are most vulnerable to the impacts of global warming (UNEP, 1989). As per the Fourth Assessment Report of the Intergovernmental Panel on Climate Change (IPCC, 2007), even a temperature rise of 1–2.5 °C will have serious effects, including increase in the severity and frequency of extreme weather, changes in rainfall patterns leading to severe water shortages and/or flooding, melting of glaciers result in flooding and soil erosion, shifts in crop growing seasons and reduced crop yields, spread of climate sensitive diseases and, an increased risk of extinction of plant and animal species. Further, sea-level rise will lead to the destruction of coasts worldwide with some islands possibly facing complete inundation. It has been estimated that about 0.41% (5,763 km^2) of the area and 4.6% (7.1 million) population of the coastal states/islands of India could be affected, if sea-level rises to 1m (TERI, 1996).

In order to alleviate the impacts of climate change, concerted action to adapt and mitigate the effects as a step of our preparedness is therefore, particularly important for coastal and island ecosystems.

IMPACT ANALYSIS ON ISLAND ECOSYSTEM

Elevated CO_2 and Temperature

Most important crops in coastal India are plantation crops, rice and vegetables which respond favourably to increase in CO_2 leading to increase in yield. Greater concentrations of CO_2 generally result in higher photosynthesis rates and may also reduce water losses from plants. Photosynthesis is enhanced when additional carbon is available for assimilation and so crop yields generally rise. The actual response to increased CO_2 differs among crops. Considering a 15% projected increase in intensity of tropical cyclones would significantly enhance the vulnerability of shallow-rooted fruit and plantation crops to the detriment of productivity.

Indirect Effects of Climate Change

Changing water demands for irrigation: Rise in temperatures may result in faster development of crops reducing crop-growing period and thereby water demand. It is likely that shortening of the crop-growing period may reduce enough plant water-use to compensate for increased water losses from plants and soils due to higher temperatures. Further, a change in field-level climate may alter the need and timing of irrigation. Increased dryness may lead to increased demand, but demand could be reduced if soil moisture content rises at critical times of the year. Because, frequent storms with higher intensity rainfall may increase precipitation to augment water supply in islands.

Surface water quality: Soil erosion and excess nutrient run-off from crop and livestock in a drainage basin may affect the surface-water quality. Changes in future farm practices, such as better matching of the timing of plant need for fertilizer with the timing of application, could possibly help to reduce the projected impacts.

Pesticide use: Pesticide use is likely to increase for most crops under the climate change conditions.

***Mangroves*:** habitats would be more vulnerable to climatic changes and resultant sea-level rise because of their unique location at the interface of the sea. By altering eco-biological processes, the inter-tidal and supra-tidal zones may extend further inland, resulting in changes in the existing ecological set-up (Jagtap and Nagle, 2007). The extent and composition of mangroves in India (4,871 km^2) may undergo major change, depending on the rate of climate change and anthropogenic activities, but the large extent of inter-tidal mudflats of 23,620 km^2 in the country may provide a scope of adjustment and adaptation in some areas, mostly in the semi-arid region. It is expected that the diversity in mangroves may improve at higher latitudes like the Gulf of Kutch; latitudinal range extension may occur at the expense of salt marsh communities; adaptation and survival chance of mangroves in deltaic region like Sundarbans will be higher than mangroves on Andaman and Nicobar Islands.

Aberrant Weather and Extreme Events

- The Andaman and Nicobar Islands suffered a severe loss of lives and means of livelihood caused due to tsunami that hit in December 2004. Nearly, 2,177 ha paddy land and 5,891 ha plantation crop was damaged (A&N Admn, 2007). The affected lands were categorized as category I: where sea water entered once and receded afterwards; category II: where sea water is entering with every high tides and category; III: where sea water entered permanently due to land subduction. The effect tsunami on affected lands could be visualized as classical example of likely scenario that may arise due to climate change. These affected lands should be considered as learning laboratory to study the effect of adaptation strategy as future action plan in coastal areas to meet the challenges at national level.

Adaptation Measures for Climate Change and Global Warming

The wide uncertainties in climate scenarios, regional variation in climate effects, and interactions of environment, economics, and farm policy suggest that there are no simple and widely applicable adaptation prescriptions. Possible adaptation measures are given below:

- ***Change in sowing dates:*** Growing of two crops instead of one with a short fallow in between to avoid excessive heat and drought in mid-summer. Winter cropping could possibly become more productive than summer cropping.
- ***Develop new crop varieties:*** The genetic base is very broad for many crops, and biotechnology offers new potential for introducing salt tolerance, pest resistance, and general improvements in crop yield and quality.
- ***Improvement in water supply, irrigation and drainage systems:*** Technologies and management methods exist to increase irrigation efficiency and reduce problems of soil degradation, but in many areas, the economic incentives to reduce wasteful practices do not exist. Increased precipitation and more intense precipitation will likely mean that some areas will need to increase their use of drainage systems to avoid flooding and water-logging of soils.
- ***Conservation tillage and crop residue incorporation:*** A warmer climate will speed the decay of soil organic matter by bacteria and fungi. Loss of organic matter reduces the capacity of soils to store water and nutrients essential for plant growth. Tillage practices that incorporate crop residues in the soils would likely combat this loss and improve soil quality.
- ***Early warning weather prediction systems:*** Accurate six-month to one-year forecasts could possibly reduce losses due to weather variability. For example, predictions of El Niño events have proven useful in regions where El Niño strongly affects weather. Conservation of biodiversity of the island.
- ***Implementation of climate change action plan for islands***: Islands entirely depend on rainfall for fresh water, which is severely scarce during four months of dry period and rationed for domestic purposes and not available for irrigation purposes. Any change in the pattern and quantum of rainfall, rise in temperature and sea-level as a consequence of climatic change would lead to scarcity/excess of water in the form of frequent droughts/ floods resulting damages to land, crops, animals affecting livelihood options in the islands. The coastal regions, particularly the island ecosystem are highly sensitive to climate

change. There is an urgent need to prepare an effective action plan after thoroughly analyzing the impacts particularly on changing patterns of rainfall, i.e., spatial and temporal variation and its impact on run-off and aquifer recharge pattern; sea-level rise; sea-water intrusions into coastal aquifers.

Mitigation Measures to Reduce Emission of CO2/GHG and Carbon Sequestration Potential of Crops

- Promoting Plantation and fruit trees in island have high carbon sequestration potential.
- Resource conservation techniques like minimum tillage, optimization of irrigation and fertilizer use.
- Organic farming.

FUTURE STRATEGIES FOR RESEARCH

- Impact analysis of climate change on soil quality, water availability and productivity of fruits, plantation and other crops, fisheries and livestock animals.
- Evaluation of suitable plantation, fruit and crop varieties under higher temperature and CO_2 levels.
- Evaluation of various adaptation strategies (crop management) for their efficacy.
- Effect of increase in sea surface temperature/acidification on marine fisheries.
- Study on bleaching incidents due to increased temperature on coral reefs.
- Analysis of sea-level rise and its effect on biodiversity, natural resources, horticultural crops, livestock and fisheries.
- Assessment of loss of land and damage to field and plantation crops, and mangrove species due to sea level rise.
- Changes in fresh water resources, saline water intrusion in to coastal aquifers and reduction in cultivated land area as a result of sealevel rise .
- Identification of suitable mangrove species for higher sea level conditions.
- Effect of water quality on shrimp farming.
- Varietal development programme for deep rooted fruit and plantation crops.
- Evaluation of sub-surface drip irrigation system for deeper root development and efficient water use in fruit and plantation crops.
- Mangroves and coral reefs provide natural protection to coasts by dissipating wave energy
- Impact of higher wave action on cage culture may be studied.

CONCLUSION

Climate change is emerging as one of the main challenges that mankind will have to face for many years to come. It could become a major threat to world food security, as it has a strong impact on food production, access and distribution. Abnormal changes in air temperature and rainfall and the increasing frequency and intensity of drought and floods have long-term implications for viability and productivity of world agro-ecosystems. It is important at this stage to visualise the likely consequence of climate change on agriculture and allied sectors and initiate research programmes that may help in sustaining agriculture and food security even under the extreme climate change stress. To devise appropriate and effective response strategies, it is essential to make vulnerability assessment studies at regional and local levels. Efforts should be initiated to protect coastal and island ecosystems from erosion, salt water intrusion using cost effective methods.

References

A&N Admn. 2007. National Conference on Agriculture for Kharif Campaign 2007-08. Department Of Agriculture, Andaman and Nicobar Administration. Presented at NASC, New Delhi, 3-4 April 2007.

IPCC. 2007. Climate Change 2007: Fourth Assessment Report of the Intergovernmental Panel on Climate Change (IPCC), WMO, UNEP.

Jagtap, T.G. and Nagle, V.L. 2007. Response and Adaptability of Mangrove Habitats from the Indian Subcontinent to Changing Climate. AMBIO: *A Journal of the Human Environment* **36:**328-34.

TERI. 1996. The economic impact of one metre sea level rise on Indian coastline-Methods and case studies, Report submitted to the Ford Foundation.

UN. 2007. The Millennium Development Goals Report. United Nations, New York.

UNEP. 1989. Criteria for Assessing Vulnerability to Sea Level Rise: A Global Inventory to High Risk Area, Delft Hydraulics, Delft, The Netherlands, 51pp.

Annexure – I

List of Important Websites for Information on Climate Change

1. www.ipcc.ch
2. http://unfccc.int
3. www.teriin.org
4. www.gefweb.org
5. www.climateark.org
6. www.unep.org
7. www.cgiar.org
8. www.worldbank.org/climatechange/
9. www.china-climate-adapt.org
10. www.climate.org
11. www.pewclimate.org
12. www.epa.gov
13. www.fao.org
14. www.sei.se
15. www.ecosustainable.com.au
16. www.greenpeace.org
17. www.metoffice.gov.uk/climatechange
18. www.wri.org
19. www.greenfacts.org
20. http://ncseonline.org
21. http://gcmd.gsfc.nasa.gov/
22. www.worldchanging.com
23. www.ecn.ac.uk.
24. www.gcrio.org
25. http://ec.europa.eu/
26. www.iisd.org
27. www.agrometeorology.org
28. www.aciar.gov.au
29. www.newscientist.com/topic/climate-change
30. www.exploratorium.edu
31. www.envfor.nic.in
32. www.climateandfarming.org
33. www.climatescience.gov
34. www.amity.edu/aigwes

Annexure – II

Research Institutes Involved with Climate Change

1. **The Energy and Resources Institute**
   Darbari Seth Block, 1HC Complex,
   Lodhi Road, New Delhi – 110 003, India
2. **World Resources Institute**
   10G Street NE Suite 800
   Washington, DC 20002, USA
3. **International Institute for Sustainable Development**
   161 Portage Avenue East
   6th Floor, Winnipeg, Manitoba, Canada R3B OY4
4. **International Institute for environment and Development**
   3 Endsleigh Street, London WC1H ODD, U.K.
5. **Climate Institute**
   900 17th St. NW, Suite 700,
   Washington DC 20006, USA
6. **Indian Institute of Tropical Meteorology**
   Dr. Homi Bhabha Road
   Pashan, Pune – 411 008, India
7. **Research Institute for Climate Change and Sustainability**
   The University of Adelaide
   Australia
8. **International Food Policy Research Institute**
   2033 Kst, NW
   Washington, DC 2006-1002, USA
9. **Stockholm Environment Institute**
   Kraftriket 2B, Stockholm
   Sweden, SE 10691
10. **Amity Institute of Global Warming and Ecological Studies**
    Amity University, D-Block, 2nd Floor
    Sector 125, Noida – 201 303
    India

Annexure – III

Important Scientific Journals on Climate Change

- Climate Policy Journal
- Climate Change
- Journal of Climatic Change
- International Journal of Climate Change Strategies and Management
- The International Journal of Climate Change: Impacts and Responses
- International Journal of Global Warming
- Asia Pacific Journal of Climate Change
- World Resource Review
- International Journal of Sustainability Communication
- Earth interactions – online journal

Annexure – IV

Important Conferences on Climate Change and their Significance

International

1. ***United Nations Conference on Human Environment, 16 June 1972, Stockholm, Sweden. It is famous as Stockholm Summit:*** This was a historic event highlighting the need for concerted effort for research and analysis of environmental issues. The conference considered the need for common outlook and for common principles to inspire and guide the peoples of the world in the preservation and enhancement of the human environment. The conference made historic declarations of guiding principles and suggested action plan for the environment by the peoples and governments.
2. ***The United Nations Conference on Environment and Development, 3–14 June 1992, Rio de Janeiro, Brazil. It is famous as Rio Summit:*** The conference reaffirmed the Stockholm declaration and resolved to build upon it, with goal of establishing a new and equitable global partnership through the creation of new level of cooperation among member states, key sectors of societies and the people. The integral and independent nature of earth, the human abode was recognized for working towards international agreement, which respects the interest of all and protects the integrity of the global environment and developmental systems. The conference proved to be a turning point for accelerated research on climate change and global warming on variety of themes and issues.
3. ***United Nations Climate Change Conference, 1–10 December 1997, Kyoto, Japan. It is famous for KYOTO PROTOCOL international agreement signed on 11 December, 1997:*** Prior to this conference in March, 1994 an international treaty the United Nations Framework Convention on Climate Change (UNFCCC) was joined by 192 countries to begin to consider what can be done to reduce global warming and to cope with whatever temperature increases are inevitable. At this conference a number of nations approved an addition to the UNFCCC treaty: the KYOTO PROTOCOL, which has more powerful (and legally binding) measures. The major distinction between the UNFCCC treaty and the protocol is that while the convention encouraged industrialized nations to stabilize the green house gas (GHG) emissions, the protocol commits them to do so. The protocol was signed on 11 December 1997 after the culmination of deliberations and negotiations in this conference.

The KYOTO PROTOCOL is an international treaty that sets binding targets for 37 industrialized countries and the European Community for reducing GHG emissions. These, amount to an average of 5% against 1990 levels over the five year period 2008–12. The protocol has been ratified by 184 countries of the convention. Further, under the UNFCCC the countries must meet their targets primarily through national measures. The KYOTO PROTOCOL offers additional means for meeting targets by way of three market based mechanisms, viz., Emission trading (the carbon market), Clean Development Mechanisms (CDM) and Joint Implementation (JI). These additional measures were the most significant path breaking achievements of the conference.

4. ***The United Nations Climate Change Conference, 3–14 December 2007, Bali, Indonesia. It is famous for Bali Road Map:*** The conference was attended by 10,000 participants from 180 countries representing government and non government organizations, press and media. It is famous for adoption of Bali Road Map, which consists of several forward-looking, decisions that represents the various tracks that are essential to reaching a secure future climate. The Bali Road Map includes the Bali Action plan, which charts the course for a new negotiation process designed to tackle climate change. An adaptation fund was launched and the scope and content of Article 9 of KYOTO PROTOCOL was reviewed. It also took decisions on technology transfer and on reducing emissions from deforestation. The most significant achievement of the conference was that the countries recognized the urgency of action on climate change and have now agreed for political response to what scientists have been telling to do.

5. ***The United Nations Climate Change Conference, 7–18 December, 2009, Copenhagen, Denmark. It is famous for Copenhagen Accord.*** The conference was attended by 115 Heads of States and several thousand participants from member countries representing government and non government organizations, press and media. The significance of the conference lies in demonstration of strong political agreement and will to cap temperature rise, reduce emissions and most importantly to raise huge finances required for the purpose. The summit ended with signing of landmark "***Copenhagen Accord***" by majority of the member states including amongst them the biggest and the richest and the smallest and most vulnerable. The accord is though not legally binding, but it was also resolved to make it legally binding in near future by consensus. The Accord strongly recognized the climate change as the greatest challenge of our time. It underlined the political will to combat climate change with common principles, but differentiated responsibilities of nations commensurate with respective capabilities. It resolved to a targeted approach to limit rise in global temperature below 2 ^{o}C on the basis of equity in the context of sustainable development and long term cooperative action. Deep cuts in emission of Green House Gases (GHG) were agreed upon and establishment of "***Copenhagen Green Climate Fund***" was initiated. A ***Technology Mission*** was established to accelerate technology development and transfer to support adaptation and mitigation measures by the member states.

National

1. ***National Conference on Climate Change and Indian Agriculture, 12–15 October 2007, NASC complex, DP Shastri Marg, New Delhi:*** This was first conference of its kind to consider the issues of Indian agriculture in relation to impending climatic variability particularly global warming affecting crop production. Importance of the conference lies in the fact that 60–70% of the population in India depends directly or indirectly on agriculture for livelihood. Crop production is estimated to decrease by 10-40% due to global warming, unless timely adaptation measures are taken. The conference recognized the urgent need for accelerated research on impact assessment, adaptation and mitigation of climate change to insulate agricultural production as for as possible from abiotic and biotic stresses caused by climate change.

2. ***Global Climate Change and Indian Horticulture: Current Status and Future Priorities, National Workshop, 6–7 September 2008, Central Potato Research Institute, Shimla:*** Indian horticulture nurtures a wide variety of fruit, plantation, vegetables, ornamental, medicinal and aromatic crops, providing direct and indirect means of livelihood to millions of people. Apart from this horticultural crops are valuable source of export earnings. Climate change and global warming is likely to have serious negative impacts on horticultural crops particularly through enhanced heat and water stress. The conference identified likely impacts on horticultural crops, research gaps and suggested future research priorities for adaptation and mitigation of climate change. Concrete recommendation for policy setting and capacity building in horticultural crops were made to take advantages of emerging opportunities from climate change. An action plan was also formulated.

3. ***National Conference on Climate Change and Clean Development Mechanisms: Opportunities & Challenges, 13 August, 2009, India International Centre, New Delhi:*** In the aftermath of the KYOTO PROTOCOL international treaty, providing for market based clean development mechanisms (CDM), many opportunities have emerged to earn carbon credits for mitigation of climate change. The mechanism stimulates sustainable development and emission reduction. India's CDM potential, including agricultural and horticultural projects in addition to industry, represents a significant proportion of the global CDM market. Emission reduction projects now swwcan earn saleable certified emission reduction (CER) credits, each equivalent to one tonne of CO_2, which can be counted towards meeting KYOTO targets. The Agricultural Finance Coswwrporation Limited organized this conference to explore various means and methods to harness this huge untapped opportunity.

Annexure – V

Proceedings and Recommendations of Brain Storming Session held at CPRI, Shimla on 6-7 September, 2008

Diversified nature of India nurtured a wide number of horticultural crops, which provide the nutritional security to millions of population. Horticulture contributes 30% of India's agricultural GDP from 8.5% of cropped area. Apart from this, horticultural crops are important from export point of view. Production and quality of horticultural crops are largely dependent on climate. Several reports (IPCC-07 and UNDP-HDR, 2007/08) on climate change projected an increase of 1.8 to 4 °C in temperature by the end of century and increase in CO_2 levels. Climate change is evident with increase in frequency of occurrence of hot days, hot nights, heat and cold waves. Climate change is likely to impact the human beings, biodiversity, forests, agriculture including horticulture. While most of the crops are likely to benefit from increase in CO_2 but an increase in temperature and water scarcity may adversely affect agriculture production thus challenging the food and nutritional security. Most of the horticultural crops are grown in ecologically sensitive zones making them vulnerable to climate change impacts. However, one can overcome these limitations by adaptation strategies. In fact, grape, a temperate fruit crop is successfully grown in tropical environments with management of plant architecture and production system. Thus it is important to study the impacts of climate change on horticultural crops, developing short and long term adaptation strategies to reduce the vulnerability and to explore capacity of horticultural crops as mitigation option.

Recognizing that the climate change is a reality impacting the production of agriculture including horticultural crops, a brain storming session on impact assessment of climate change for research priority planning in horticultural crops was organized at CPRI, Shimla on 6-7 September, 2008.

While inaugurating the brain storming session Hon'ble Governor of Himachal Pradesh Her Excellency Smt. Prabha Rau highlighted the importance of adaptation strategies in major developmental activities, new land use arrangements identified by science and technology and financial incentives and package for improved land management for developing and adopting Green House Gases inhibiting technologies. She mooted the idea of establishing a Green Research Fund for strengthening research on adaptation, mitigation and impact assessment. On this occasion the Hon'ble Governor of Himachal Pradesh released a book on 'Impact assessment of climate change for research priority, planning in Horticultural crops published by CPRI, Shimla. In his key note address Dr. HP Singh, DDG (Horticulture), ICAR provided

Dr H.P. Singh, DDG (Horticulture) addressing the delegates

Her Excellency Smt. Prabha Rau, Hon'ble Governor of Himachal Pradesh, addressing the Inaugural function

an overview of climate change, its impacts and also adaptation and mitigation strategies and emphasized upon research needs in short and long term to face the future challenges including assessment of climate change impacts on horticultural crops and development of adaptation strategies. He urged to address i) priority of education, research and development and policy implications for enhancing adaptive capacity of Indian horticulture to climate change, ii) capacity assessment of Indian horticulture for mitigation of GHG, iii) appropriate short term and long term action plans to mitigate the impact of climate change in horticulture. In technical sessions chaired by Dr. HP Singh, DDG (H), ICAR, scientists presented the studies on impact of climate change on different crops and pathogens. The highlights are as follows:

Observed impacts of climate change related events on Indian horticulture

Temperate fruits

- Warmer climate and non-fulfillment of chilling requirements during winter in lower elevations up to 1500 m above msl resulted in decline of apple productivity to the tune of 40–50%. This resulted into shifting of apple production to higher elevation (about 2700 m amsl). Apart from this warm temperatures during autumn, just before harvesting, lead to increased frequency of sunburn and poor development of colour affecting the marketability.
- Mild winter temperatures followed by warmer springs advanced bud burst exposing buds to frost damage in almond and apricot.
- Elevated CO_2 conditions increased the content of aromatic compounds in strawberry fruits.
- High temperature and moisture stress resulted in sun burn and cracking in apples, apricot and cherries.

Tree fruit crops

- In mango, high temperature and moisture stress resulted in dehydration injury to panicles and low fruit set. On the other hand, low temperature (4–11.5 °C), high humidity (>80%) and cloudy weather in January delayed panicle emergence. Apart from this low temperatures during inflorescence development reduced number of perfect flowers.

- In citrus, untimely winter rains promoted vegetative flushes instead of flowering flushes. Dry spell during flower emergence and fruit set affected flower initiation in citrus and aggravated pest (*Psylla*) incidence.

Tropical fruit crops

- Banana leaf production increased with increase of 1-2 °C temperature beyond 25–30 °C thereby reducing crop duration and increasing production. Increase in temperature with low RH (less than 60%) reduced leaf spot diseases in banana especially in sub humid region.
- In vine grapes anthocyanin development is influenced by difference between day and night temperatures with high variation (15–20 °C) promoting colour development.

Vegetable crops

- In potato, frost damage reduced tuber yield by 10–50% depending upon intensity and coincidence with sensitive stage, while yield loss due to high temperature was to the tune of 10–20% depending upon coincidence with sensitive stage. High temperatures reduced marketable grade potato tuber yield to the extent of 10–20%. With increase of 1°C in ambient temperature, appearance of potato peach aphids advanced by two weeks which reduced growing period for seed crop in potato.
- In onion, temperatures above 40 °C reduced the bulb size, while elevated CO_2 is beneficial in improving bulb size and yield. In cauliflower, curd initiation was delayed due to increased temperature.
- In tomato, temperature above 30 °C induced maximum flower and fruit drop and high temperatures after pollen release decreased fruit set and fruit yield, while elevated CO_2 (550 ppm) enhanced tomato production by 24%.

Arid region horticulture and spices

- Occurrence of frost during January in Rajasthan has affected cumin resulting in total crop failure.
- Frost injury was also observed in young budded plants of aonla, where scion portion is affected resulting in dried young grafts. In late maturing varieties, due to low temperature and frost fruits become whitish, water oozes out subsequent to which they dry and turn back.
- In pomegranate, frost injury caused hardened fruits with dull skin colour and rot making them unmarketable.
- In black pepper, increase in rainfall during December-January decreased the productivity, while March-April rainfall increased productivity.

Plantation crops

- In coconut, impact of climate change related events like consecutive droughts (in TN and Karnataka) and cyclones (in AP) adversely affected nut yields. Consecutive droughts killed lakhs of palms and farmers adapted by shifting the cropping pattern to pulses and millets. Studies on response of coconut seedlings to elevated CO_2 (550 and 700 ppm) indicated increased growth of seedlings by ~20–30%. Development of coconut simulation model enabled to study the climate change impacts in future scenarios. Projections indicate that climate change impact on coconut yield, simulated using coconut simulation model for HadCM3 A2a, B2a and A1F 2020, 2050 and 2080 scenarios, has positive effect in west coast and parts of TN and Karnataka (provided water availability exists at current level) and negative effect in east coast, parts of TN and Karnataka. Studies on the carbon sequestration potential of coconut indicated that annual carbon sequestration in coconut above ground biomass varied from 15 Mg CO_2/ha/year to 35 Mg CO_2/ha/year depending on cultivar, agro-climatic zone, soil type and management, out of which 2-3 Mg CO_2/ha/year stocked into stem. The carbon sequestration potential of arecanut plantations ranged from 5.14–10.94 Mg CO_2/ha/year while that of cocoa plantation ranged from 2.02–3.89 Mg CO_2/ha/year making all three of them strong candidates for carbon trade under Clean Development Mechanism.

- In cashew, heavy rains during harvest period adversely affected production and quality of cashew nuts due to germination. Extension of monsoon rains beyond October in west coast delayed bud break and flowering. The tea mosquito bug populations and damage score in cashew were negatively correlated with minimum temperature (15–20 °C), RH (85–90%) and positively with bright sunshine hours (9–10 h). Bright sunshine hours less than 2 h per day triggered incidence of inflorescence blight.

Tuber crops and others

- At elevated CO_2 (700 ppm) dry weight of cassava tubers increased eight fold while WUE was almost doubled.
- Quality (carotene, starch and glucose content) and tuber yield of sweet potatoes increased in elevated CO_2.
- Increased temperature beyond 20 °C during winter affects cultivation of seasonal button mushroom and increase incidence of insect pest and diseases.
- Temperature rise from 20 to 22 °C increased the incidence of pest and diseases in case of *Cymbadium* orchid.

Research gaps identified

The following research gaps were identified.

1. Information on past climate trends in areas of target crop (analysis should be done for at least 30 years period) is lacking except for a few crops.
2. The relationship between variations in observed weather and productivity of target crop is to be established in most of the crops.
3. Lack of simulation models for most of horticultural crops hampers the researchers to project the impacts in future climate change scenarios.
4. Very few studies are being carried out on response of horticultural crops to elevated CO_2 and temperature.
5. Only a few studies are conducted on impact of climate change related events on quality of horticultural crops.
6. Lack of information on socio-economic impacts of climate change on target horticultural systems.

7. Information need to be generated on carbon sequestration potential of horticultural crops particularly of perennial crops for possible exploitation in CDM.
8. More eco-friendly adaptive strategies need be generated for reducing the impacts of climate change.

Future research priorities

a) Impact assessment

Quantification of climate change impacts on target crops is a vital step for developing adaptation strategies to minimize the vulnerability or to maximize the positive impacts. For this several methodologies are adopted such as analysis of past weather and yield data, use of simulation models and meta analysis. In order to assess the response of plants to elevated CO_2 and temperature, experimental approaches used include growing plants in Open Top Chambers (OTC), Free Air Carbon Enrichment (FACE) rings, Free Air Temperature Enrichment (FATE) rings and phytotron facilities.

In impact assessment the study areas include:

Analysis of observed climate impacts on horticultural crops such as potato, vegetable crops, garlic, fruit crops (mango, citrus, litchi, banana), medicinal and aromatic pants, tuber crops (cassava and sweet potato) and plantation crops (cashew, oil palm) with reference to increased temperature/extreme events.

Monitoring phenology of perennial crops such as tree fruit crops, plantation crops as bio-indicators of climatic variability and change.

Development of simulation models for major horticultural crops. Currently simulation models are available for only a few horticultural crops such as potato, coconut, tomato, onion, banana, *etc*. Development of simulation models for other horticultural crops such as vegetable crops, fruit trees (apple, mango), plantation crops (cocoa, oil palm, cashew) and tuber crops (cassava and sweet potato) should be a priority area of research.

Studies on response of horticultural crops to elevated CO_2 and temperature. Such studies should be conducted on crops such as apple, banana, potato, vegetables, cassava, sweet potato, coconut, arecanut, cocoa, spices (black pepper) *etc*., Identification of climatic thresholds having impact on yield and quality needs emphasis. Also studies on impact of climate change should include on quality of pre and post harvest produce in fruit trees (grapes), vegetable crops, spices, plantation crops, secondary metabolites of medicinal and aromatic plants. Use of inundated land for studying the impacts of sea-level rise on Andaman Nicobar Island ecosystem.

b) Adaptation

Adaptation strategies to climate change include change in sowing dates, development of cultivars tolerant to abiotic stresses such as heat, cold, *etc*, management of plant architecture, development or refinement of eco-friendly adaptive strategies for minimizing the adverse impacts of climate change or for maximizing the benefits of climate change.

Development of abiotic stress tolerant cultivars: Development of cultivars with low chilling requirement in temperate fruits including apple; banana varieties to with-stand excess rainfall; mushroom varieties with prolonged shelf life, and moisture stress and heat stress tolerance; varieties responsive to pruning and cool night low requirement for colour development in guava; development of ideal root stocks against biotic and abiotic stresses in citrus; short duration, heat, drought and salinity tolerant cultivars in potato, seed spices, onion, tomato, garlic, mango; frost tolerant varieties in arid region horticultural crops, seed spices, cassava and sweet potato; drought tolerant and compact palms in oil palm; varietal development for deep rooted fruit and plantation crops in island ecosystems. Identification, conservation and utilization of usable gene pool in orchids with reference to climate change impacts.

Management of plant architecture to minimize the climate change impacts on tree fruit crops (apple, mango, grapes, citrus), plantation crops and medicinal tree plants.

Development/refinement of adaptation eco-friendly agro-technologies for minimizing the climate change impacts and maximizing the climate change related benefits, if any, including water and fertilizer management (mulching, drip irrigation, fertigation, specific pre-harvest managements like bagging of grape bunch, sub-surface drip irrigation, pulse irrigation) in fruit trees (apple, mango, grapes, citrus), potato, vegetable crops (onion, tomato, garlic), arid region horticultural crops, seed spices, orchids, spices, plantation crops (coconut, arecanut, cocoa, cashew, oil palm) and medicinal and aromatic plants.

Assessment of impact of climate change on pests and diseases of horticultural crops and development of adaptation measures: Studies on dynamics of pest and dieses in relation to climate change scenarios in potato, arid region horticultural crops, banana (leaf spot diseases and virus infections), spices, seed spices, development of early warning system for pest management; development of biotypes/strains of natural enemies of pests under high temperature regimes; management of TMB and inflorescence blight in cashew.

Analysis of impacts on - and adaptation capacity of target socio-economic systems includes studies on adaptation capabilities of farmers, climate change driven changes in cropping patterns.

Development of compendium of ITK and refining them for better exploitation. Traditional Indian agriculture is a repository of several cultural practices which are eco-friendly and sustainable. Hence efforts should be made to document them reasoning in light of adaptation to climate change impacts.

Develop new land use systems for better integration of natural resources to exploit climate change related changes.

Development of weather based advisory services for short, medium and long range forecasting. Development of pest and disease forecasting systems forms vital part of crop management strategies.

c) Mitigation

Mitigation is referred to the process in which the emission of green house gases are reduced or they are sequestered. Development/refinement of GHG emission inventories is one of the important areas of mitigation research. Improved agronomic practices for enhanced nutrient (nitrogen, inorganics and organics) use efficiency for reduction of GHGs and eco-friendly disease and pest management strategies also form mitigation options.

Estimation of carbon sequestration potential of different horticultural based land use systems such as perennial medicinal plants and plantation crops (coconut, cocoa, arecanut, cashew and oil palm). Apart from this, development of composting techniques to minimize emission of toxic gases in mushroom cultivation, vermicomposting are components.

Capacity building

In order to prepare Indian Horticulture to face the changes posed by climate change impacts, emphasis should be laid on ***infrastructural development and capacity building activities*** like strengthening weather stations, creation of infrastructure for climate change studies, development of crop simulation models, capacity building for carbon sequestration studies and intensify efforts for increasing climate change literacy among all stake holders of horticulture.

Developmental projects for adaptation

Apart from the above, immediate emphasis should be given to ***surveillance of pest and diseases, integrated water management, weather based insurance to farmers.*** For this it is important to understand the impact of environment on production of target crops resulting in to establishment of weather watch groups for climate sensitive commodities.

Policy setting

Policy setting includes, ***strengthening education, research and developmental activities for enhancing adaptive capacity of Indian horticulture to climate change***. Augmenting production of horticultural crops including improved land use pattern to suit future climate change scenarios. This will result in improvement and use of natural resource management. It is important to promote technologies for improved nutrient use efficiency. Future planning activities should consider incorporation of climate change impacts. Strengthening water harvesting and conservation options and promotion of water use efficient and recycling technologies require immediate attention. Incentives for eco-friendly technology adaptation and environment services provided by the farmer should encourage farmers for adapting green technologies. Exploring CDM benefits for mitigation strategies is most important. Development of comprehensive contingency plans for reducing the vulnerability and/or maximizing the benefits due to climate change (short, medium and long-term) should be given priority.

Action plan for research and assessment

Immediate action warranted is providing financial support for developing infrastructural facilities for climate change research and strengthening inter-institutional collaboration for climate change studies for better utilization of resources and expertise. It is also important to organize annual review meetings for mid-term corrections and improvements.

Author Index